A GRASSROOTS HERITAGE

Texas Farm Bureau's Six Decades of Service

By
Steve Kelton

ISBN: 1-87234-29-7

Published by Heritage Publishing, Inc., 9029 Directors Row, Dallas, Texas, 75247. Phone: 214-630-4300

TEXAS FARM BUREAU

President . . . . .Bob Stallman
Executive Director. . . . . .Vernie R. Glasson III
Director of Research, Education and Policy Development
and Project Coordinator . . . . .G. Glen Jones, Jr.
Project Assistant . . . . .Bonnie Montgomery

HERITAGE PUBLISHING

Publisher . . . . .Rodney L. Dockery
Editorial Director . . . . .Caleb Pirtle III
Executive Editor . . . . .Kenneth E. Lively
Production Manager . . . . .Blake Judkins
Art Director . . . . .Janet Todd

Texas Farm Bureau and Heritage Publishing would like to thank the Texas Department of Agriculture and the Texas Cotton Ginners' Association for the use of their photography in this book.

Printed by The Bidal Group: David Terrazas, president

Manufactured in the United States of America

First Printing

FOREWORD

Since the beginning of recorded history, we have studied the past. Countless school children have asked, "Why? What possible significance can this have in my life?"

To paraphrase one scholar, we must study history or we are doomed to repeat it. Of course, there are many things about our history as a nation, a state and as an organization that are worth repeating. Perhaps that is another reason to recall the past.

In my own view, our past, our traditions and our actions are the building blocks upon which our culture and lifestyle rest. We study history to understand ourselves. We are, ultimately, the product of what has happened before.

On the pages of this book, you will read the first historical account of the Texas Farm Bureau that has been commissioned by the organization. You will read of the great accomplishments of Texas Farm Bureau, but also of the mistakes. You will learn about the past leaders of this organization who, against great odds, created a movement that has stood the test of time. It is a story of courage, tradition and vision, and of the obstacles that stood in the way.

This book is an overview of the organization, showing how Texas Farm Bureau has improved the agricultural industry and dealt with the important issues that agriculture has faced. It is a "who we are and what we have done" book.

Our leaders did things no other farm organization had been able to do. Today, no other farm group has been quite as successful as Farm Bureau. We should not forget the struggle, the difficulty in creating such an organization. It was built from pioneer stock, somehow managing to combine the strengths of individual freedom and the power of numbers.

Texas Farm Bureau is one of the Lone Star State's great success stories. It is a story of rural people who often felt isolated, unappreciated and ignored by the powers that ran Texas. They so much desired a voice in what happened to them that they created their own.

Our organization, so strong today, is the result of that need to have an impact on the laws, regulations and social actions that affected them. Our Farm Bureau is the product of the dreams for a better life that inspired our pioneers.

From the first halting steps of an infant organization, to the political and economic power that Texas Farm Bureau has become today, you will enjoy the accounts in this book. If there are lessons to be learned from our own history, we went to great lengths to include all of them in this volume. The reading of this book will help us understand more about our organization and our members who were, and are, the real strength of Texas Farm Bureau.

—Bob Stallman
President, Texas Farm Bureau

DEDICATION

To all those who have
served and
continue to serve
Texas Agriculture

PREFACE

The Texas Farm Bureau is the greatest example of a grass-roots organization in America. No other membership organization offers as much opportunity for individual member involvement—from youth activities, to determining public policy positions, to serving in leadership roles. With a proud history of accomplishment for farmers, ranchers and other rural Texans, Farm Bureau is well prepared to face the future challenges to agriculture that undoubtedly will come.

The Texas Farm Bureau is a general farm organization, representing agricultural producers of virtually every commodity known to grow in this big old state—from grapefruit to alfalfa, beef cattle to sesame, milo to timber, Angora goats to potatoes, honey bees to cotton, to name only a few! And Texas Farm Bureau has been there for these producers, working for every major legislative or regulatory battle affecting agriculture during the past sixty years.

We are there on other fronts as well. The battle for public opinion has always been a target for Texas Farm Bureau, as has the battle for the support of elected officials. On the economic front, Texas Farm Bureau has developed the widest array of service and benefit programs that one can find in membership organizations.

Insurance services, financial services, group purchasing programs and numerous others provide a win-win opportunity for members to save money and for the Texas Farm Bureau to help fund its organization activities.

The Texas Farm Bureau is, without question, the leading agricultural organization in the Lone Star State. It has long been the number one lobbying organization representing farmers and ranchers with the Texas Legislature and Texas Congressional members. It has more recently become a major political force—with one of the most active and aggressive non-partisan political programs in the state.

As a Texas Farm Bureau member—or as a prospective member—the story that this book tells should make you proud.

—Vernie R. Glasson
Executive Director
Texas Farm Bureau

Contents

Texas Farm Bureau: The Early Years

The Texas Farm Bureau, today the largest and most influential agriculture organization in the state of Texas, is not the first entity to bear that name. Phoenix-like, the current organization rose from the ashes of a previous Farm Bureau more than sixty years ago.

In early December of 1933, approximately six hundred delegates to the Texas Farm Bureau Federation's annual convention in Dallas voted that organization out of existence and replaced it with a new one. The new association was organized by the same people who read the old one its last rites, and its treasury was seeded with the old Farm Bureau's surviving funds, but the two were entirely separate, even to their names. The new organization did not adopt the title "Texas Farm Bureau" for several more years.

That wasn't long enough for some people. Many producers remembered the old Texas Farm Bureau Federation mostly in relation to a cotton marketing cooperative that it spawned and which essentially devoured it. The ambitious co-op was no stranger to controversy, and even though the new organization had nothing to do with the old, the ghost of the earlier arrangement would haunt organizing and membership efforts of the new Farm Bureau for decades.

For that reason, the story of today's Texas Farm Bureau cannot properly begin without an understanding of what came before it.

The Farm Bureau movement nationwide had its earliest roots in Texas, beginning shortly after the turn of the century. Unlike many agriculture organizations both before and after — from the Grange to the Farmers Alliance, to the Farmers Union and the much more recent American Agriculture Movement, the Farm Bureau movement did not originate as a protest. Its first impetus was to educate producers.

Sophisticates have long enjoyed portraying Texas as a backwater — long on braggadocio but short on culture, education and forward thinkers. That image has persisted, though time and again, Texas and Texans have proven it wrong. It was the "backwater" of rural Texas, in this case, that spawned the "demonstration" method of agricultural education.

That, in turn, led to establishment of the cooperative federal-state agricultural extension service and the Farm Bureau movement which was formed at first to nurture it.

The year was 1903, the place Terrell, Texas, on the state's east side. As difficult as it may be to imagine today, agriculture at that time still formed the foundation of most communities, urban as well as rural. Almost half of the populace still lived on the land, and town and city commerce alike depended on rural people; there weren't enough city folk to keep the merchants in business, support the banks, and consume what goods were manufactured. Thus, when country people suffered, their city cousins shared the pain, and in 1903 the East Texas cotton patch saw serious pain just over the horizon. It took the form of a small insect, the Mexican boll weevil. Nearly a century later Texas and surrounding states would still be fighting the boll weevil and finally be zeroed in on its destruction after spending decades learning to accommodate the pest and work around it. In 1903, however, accommodation was only a dream, and eradication an even more distant hope.

The first concern for businessmen in the Terrell area was to avert the panic that they had seen in farming communities hit earlier. There was a real fear that panic among area farmers could bring the town to ruin. Facing up to that possibility, Terrell businessmen gathered in the spring of the year to map a strategy. The key proved to be Dr. Seaman A. Knapp, an agricultural educator with experience in such disparate locales as New York State, Iowa and Northern Louisiana.

More than a decade of bringing what were then "modern" farming techniques to Louisiana, producers had convinced Knapp that showing farmers how to do something new beat telling them. His allies in that work were the farmers who had already learned and benefitted; when news got around that a producer was making money on a crop that would barely break even for his neighbors, those neighbors were halfway to becoming converts.

Knapp brought his "demonstration" method of teaching to the Terrell vicinity, where the concerned captains of commerce managed to convince a local farmer — after guaranteeing him an indemnity should things go wrong — to set aside part of his land for cultivation under Knapp's direction. The demonstration succeeded, panic was averted in the Terrell area, and other farmers were soon won over to Knapp's improved farming techniques.

More important, Knapp's teaching method itself had also gained converts — converts with clout. Within the next three years, similar hands-on, real-world demonstrations were underway throughout the South, led by agents of the newly-formed Farmers' Cooperative Demonstration project, supported by businesses, banks, railroads and colleges, and underwritten by a quarter of a million dollars in federal funding.

Local governments began to join in the effort, and in 1906 Smith County in Texas became the first to contribute toward the salary of a permanent, resident "demonstrator." The term "county agent" entered the language. Other counties and states followed suit, and by 1912, the South boasted almost 860 such agents. The concept was already spreading from its roots; the 1912 tally included "collaborating" agents assigned to teach girls and women such domestic projects as canning and backyard poultry production.

The North was slower to recognize the import of the program arising in the South, but by 1910, Broome County, New York, took the first steps in transplanting the concept, adding a wrinkle that formed the basis of the Farm Bureau movement.

Centered at Binghamton, Broome County was typical of much of rural New York State at the time, dotted with abandoned farms and others that soon would be if their operators continued to employ outdated techniques. Like its counterpart in far-distant Texas, Binghamton's Chamber of Commerce recognized the need to rescue its deteriorating agricultural trade base. Chamber leaders organized a tour of Broome and surrounding counties, inviting representatives of both the New York and United States agriculture departments, as well as the state's land-grant university, the New York State College of Agriculture.

What they learned was discouraging enough to alarm regional railroad officials whose traffic depended in large measure on agricultural production. The Lackawanna Railroad pledged its support for whatever agricultural development plan might be devised, and an official of the road

The fight against the boll weevil in East Texas in 1903 began to band many farmers together.

joined two progressive farmers to form an agriculture division or "bureau" of the Chamber. The Binghamton "farm bureau" consulted with experts from both USDA and the agricultural college, with an eye toward establishing a model "demonstration farm" within the county. USDA spokesman, Dr. W.J. Spillman, recommended instead the southern approach of conducting demonstrations in cooperation with respected local farmers, and by the spring of 1911, Broome County hired the first county agent north of the Mason-Dixon line.

Funding was split between the Binghamton Chamber of Commerce, the Lackawanna Railroad and USDA, with technical assistance provided by the agricultural college. Oversight was in the hands of the "farm bureau," which soon expanded from its Chamber of Commerce base with the appointment of "community chairmen" from throughout the county.

With Broome County as an object lesson, the southern "county agent" idea quickly caught on in other counties throughout the North and Midwest. Meanwhile, the northern notion of turning control over to a "farm bureau" went south just as the county agent concept had gone north, and both regions of the country were soon on a roughly similar track — the Farm Bureau movement.

Across the country, farmers and ranchers were meeting to hear the benefits of a farm bureau movement.

Further refinements took bureau control out of the hands of banks and the other commercial institutions that tended to dominate

chambers of commerce, vesting authority instead in dues-paying farmer members. Finally, the federal Smith-Lever Act of 1914 reorganized what up to that point had been a haphazard system of partial federal funding for county agents. Under the act, federal funds were made available through the same land-grant colleges that had earlier been tapped to create and operate the experimental station program. With state and local contributions added in, the agricultural "extension" service was underway, and the farm bureau movement was in the education business.

Early farm life was a hard, demanding and unforgiving life for both men and women.

As additional county "farm bureaus" organized, it was only natural that they would lean on one another as well as the pioneering groups for advice, assistance and moral support. That led to the next logical step, the organization of county farm bureaus into state confederations.

In 1917, within six years of the creation of the Broome County Farm Bureau in New York, that state boasted thirty-nine active county farm bureaus. Thirty-four of them, representing nearly forty thousand farmers, sent delegates to a meeting that would culminate in the New York State Federation of County Farm Bureau Associations, the nation's first such statewide organization.

It is one of the little ironies of history that Texas, where county demonstration programs began, was to lag behind later starters in the programs' evolution into the farm bureau movement. By that time, farm bureaus across the country were expanding beyond their original goal of directing demonstration work; the concept of cooperative marketing had begun to share the spotlight with education, and that additional focus helped spur organizational efforts in a number of states, including Texas.

War — and its aftermath — added to the impetus. One of the Allies' slogans during World War I was "Food will win the War," and the United States government took that sentiment to heart well before the nation was formally committed as a combatant. Increased agricultural production, along with more efficient shipping and marketing, were goals assigned to the U.S. Department of Agriculture. USDA sought to meet that responsibility with improved education and organization, and it found a ready-made network of agents already on the ground in its own cooperatively funded extension-farm bureau programs. The agency put emergency money and its organizational skills into the effort of expanding the still somewhat spotty network, and several new states joined the fold during the war.

Organizational momentum was still building when the war ended,

Early-day farmers and ranchers had to battle bad weather and worse roads leading from their isolated farmsteads.

and the fruits of stepped-up wartime agricultural production were still being harvested. The consequences of the latter began to hit home before the former ran out of steam; the temporary, artificially inflated demand of the war years quickly abated with the Armistice, and farmers found themselves in a buyer's rather than a seller's market. Their difficulties were compounded for a time by fallout from similar conditions in other industries and the return of an army of doughboys who were no longer on Uncle Sam's payroll. The strain on the nation's economy and on the individual economies of working men and their families was evident in a tally of more than three thousand labor strikes during the year following the war. What people couldn't buy, the farmer couldn't sell, and what people couldn't afford to pay, the farmer couldn't charge. Agriculture was in a pinch.

Against that backdrop, twelve states sent representatives to a February 1919 meeting in Ithaca, New York, to consider organizing a *national* federation of farm bureaus incorporating state groups, just as the states had done with their counties. A USDA estimate put the number of county bureaus nationwide at seven to eight hundred by then, though only nine state federations were formally chartered. The main product of the Ithaca meeting was a structure for a larger gathering scheduled in Chicago the following November.

The fall meeting included delegates from thirty-one states; most of those, including Texas, were not yet organized, but one of the objectives of the still-unformed national association was to foster more county and state groups. Delegates were courted especially from states where organization was deemed likely with a little push.

The Texas contingent included voting delegate R.R. Bowen and Texas A&M University representatives T.O. Walton, W.B. Lanham, and Walton Peteet. Walton was A&M's extension director, Lanham his assistant director, and Peteet the extension marketing specialist.

Documentary evidence of their impressions of the national gathering no longer survive, but the quartet presumably returned from Chicago fired with organizational zeal, because Texas held its own organizing meeting the following month in Waco. Invitations bearing the signatures of Texas A&M officials and prominent farmers went out statewide, and sixty counties provided representatives. The confab gave birth to the Texas Council for Agriculture and Home Economics, presided over by Dallas County farmer John T. Orr and extension organizational specialist H.L. McKnight, who served as temporary secretary.

Working backwards, the group then began organizing at the county level. By the fall of 1920, directors voted to change the Council's name to the Texas Farm Bureau Federation. They associated with the national organization, paying dues of fifty cents per head on five hundred members. Texas' anticipated importance to the young American Farm Bureau Federation was clearly out of all proportion to its still-meager membership roll, as evidenced by the immediate elevation of Orr to the national body's executive committee and the appointment of Coleman County stockman J.E. Boog-Scott to the executive committee of the AFBF wool marketing division.

One of the first orders of business for the Texas organization was to create a cooperative marketing program for cotton. All of agriculture had been hit by the postwar slump, but cotton farmers were especially desperate. The cotton market opened in 1920 at forty cents per pound, but by the end of the year, farmers were lucky to get a dime. Cotton farmers had long been victims of a marketing system dominated by middlemen and remarkable for its record of waste, inefficiency and outright thievery. As long as prices remained acceptable, there had been no great agitation to replace traditional, if unpopular, marketing methods with anything new; the old practices were at least familiar, and anything else would be a leap into the realm of the unknown.

By late 1920, however, farmers were ready to leap. Their ten-cent cotton had cost many of them thirty cents to raise and harvest, and they were losing a hundred dollars a bale — most of it borrowed money. The Texas Farm Bureau Federation responded with a radical cooperative marketing plan along the lines of a proposal floated earlier that year in a dramatic and fortuitously timed Montgomery, Alabama, meeting. It struck such a chord with embattled cotton producers that the federation was able to absorb the Texas branch of the American Cotton Association and the United Cotton Growers of America, and even corralled the blessing of the

Texas Farm Bureau's first home in Waco was a free upstairs office of the Waco Chamber of Commerce on Franklin Avenue.

rival Farmers Union.

The plan created a state cotton marketing agency known as the Texas Farm Bureau Cotton Association, its primary purpose to raise cotton prices to farmers by eliminating middlemen, brokers and speculators, and by controlling enough cotton to deal directly with mills and even foreign buyers. Only bonafide cotton growers could be members, and they were required to sign a contract promising to "sell and deliver" to the association all of the cotton they produced for a five-year period. The association in turn would pool, grade and resell members' cotton throughout the season, and the members would later be paid based on the average price received for cotton of that given grade. The idea was to sell cotton gradually, as the market would bear; if the association could control a sizeable percentage of production, it could prevent or at least minimize untimely marketing and the consequent price fluctuations. Given enough control, it could potentially command its own price.

The cooperative never controlled as much cotton as anticipated, however. In its first year, the organization obtained nearly twenty thousand "ironclad" contracts from growers who claimed to represent 600,000 bales of cotton. That was a far cry from the one million bales upon which the association's premise of substantial market influence was predicated, and even the lesser figure turned out to be highly overstated. Unfortunately, the field men employed to sign up new members were paid on the basis of the contracts they secured, not the cotton their contractors actually delivered. In addition, growers could defer payment of dues by electing to have their ten dollar annual contribution deducted from proceeds on the cotton they delivered; if they never delivered, they didn't pay, so it cost a grower nothing to sign up. By the end of the first year, fewer than twelve thousand farmers actually delivered cotton, and the total tonnage amounted to less than ninety-four thousand pounds.

Participation increased in subsequent years, to the extent that the Texas Farm Bureau Cotton Association established its own financing corporation and even built its own gins. Little of that clout rubbed off on the American Farm Bureau Federation, however. The Texas federation was supposed to remit fifty cents of each member's dues to the national Farm Bureau, but conflicting records demonstrate that it never gave much more than lip service to that chartering agreement: TFBF, which claimed nearly twenty thousand co-op members in thirty-four counties during the founding year of 1920, reported only 2550 members to the American Farm Bureau Federation. For 1921, TFBF recorded thirty thousand members in 134 counties, but reported and paid national dues on fewer than eleven thousand.

The cooperative was clearly the focus for most members, and it gradually began to overshadow its parent entity. By the mid-Twenties, the Texas Farm Bureau Federation ceased to collect dues directly from its membership, drawing financial support instead from dues remitted by the co-op. That was the beginning of the end for the original Texas Farm Bureau, as the cotton co-op increasingly centered attention on itself, neglecting its "superfluous" parent.

As the clouds of the Great Depression rolled in, the federal government in 1929 passed a farm bill intended to shore up falling prices by using cooperatives such as the Texas cotton association to stockpile commodities. The project failed, as most stockpiling schemes ultimately do, but in the process it split the Texas Farm Bureau Federation from the cooperative, which came to be known as the Texas Cotton Growers Association. By 1932, the Texas Farm Bureau Federation was reduced to hardcore believers, reporting only a token one hundred paid memberships.

Federation directors and officers formally dissolved the fourteen-year-old organization at its annual convention in 1933, and convention delegates, without a dissenting vote, immediately rolled the defunct group's small treasury over into a new entity, the Texas Agricultural Association. They envisioned that the new statewide farm organization would have a wider scope of possibilities and be more in line with the requirements of producers to meet the problems that afflicted rural Texas as it battled to survive the Depression era.

Delegates actually made the change seeking to secure the active membership of other large farm organizations which refused to affiliate with the old federation. The newly-formed TAA was to carry on membership, social, educational and legislative activities in close cooperation with the buying and selling unit the convention delegates organized at the same time.

H.G. Lucas, a Brownwood pecan grower, was serving as president and Mrs. M.S. Hudson of Hale Center held the position of vice president when the Texas Farm Bureau Federation was gaveled out of existence. Lucas, long an active leader in the state's co-op movement, had worked long and tirelessly to hold Texas farmers and ranchers together after TFBF lost its affiliation with the American Farm Bureau, even though membership was dwindling. Rural Texas appreciated his efforts, his devotion, and Lucas, along with Mrs. Hudson, were immediately elected to take up the same posts at the helm of the new TAA.

In time, Lucas moved the organization's office to Dallas from his plantation in Brownwood. He hired a secretary and began publishing *The Farmer's Banner,* exclusively for members of TAA. It was a struggle. Membership was low, and finances virtually non-existent, but H.G. Lucas almost single-handedly kept the organization alive and functioning for the next four tough years, battling economic hard times and the devastating effects of the calamitous Dust Bowl. Officers of the new group elected not to collect membership dues directly. Instead, they sought to broaden their appeal beyond the cottonfield by soliciting participation from members of such local organizations as cooperatives, home demonstration, 4-H and FFA clubs; the TAA would draw its support from those bodies.

That soon proved to be a flawed plan, much as the idea of supporting the old TFBF on contributions from the cotton cooperative. TAA delegates met in College Station after a year of operating with a near-

An enterprising Texas farmer pinned hopes and dreams on a tractor, which certainly beat plowing with a team of mules, as he waited for the AAA program to bring parity on fair prices for his crops.

empty purse, and they brought two well-learned lessons with them. The first was that the organization had to collect dues if it was ever to operate on a sound footing. The second lesson was that TAA needed some process that would assure it was being run by its members and not the other way around.

The delegates accordingly began inching toward the structure that would later define the Texas Farm Bureau. They revamped the TAA's constitution to vest responsibility and authority for local decision-making in individual county-level entities, and they voted to levy dues. At that point the dues process was still indirect; each county organization was to be assessed twenty dollars per year, which might variously be raised from individual members or from among the constituent clubs that TAA still considered the foundation of its county groups.

Montague County native Austin Crownover would go to work for the Texas Farm Bureau more than a decade later and eventually rise to the post of state sales director for the insurance companies. At the time of the College Station meeting, however, he was a student at Texas A&M. Nearly sixty years later, he recalled that early gathering in College Station's YMCA building.

"Being an ag student and knowing about the meeting," Crownover explained, "I went to see what was taking place. I remember one of the things they agreed to do — this was when times were real hard — was to set up county organizations and charge each county ... for membership. Not individual members, county members."

Even with the improved structure, the TAA struggled; a cooperative marketing and purchasing program originally planned by the organizers fizzled, and little remained of the state group's lofty visions beyond an educational thrust — and politics.

It was, after all, during the Great Depression, and unrest stirred throughout the country, particularly in the agricultural sector. However, survival became possible for the Texas farmer when the administration of Franklin D. Roosevelt attacked the problem with the most elaborate farm program in the nation's history. To some, it seemed that the president made relief for agriculture a sort of cornerstone for his entire New Deal structure. At the heart of the plan was a concept known as parity. To the farmer, it meant fair prices. The Agricultural Adjustment Administration, was established to limit production of so-called "basic" crops, and thus raise prices, by assigning acreage limitations to individual farmers. The limitations were overseen by local "production control committees," and compliance was encouraged with benefit payments to which non-cooperating farmers were not entitled. The AAA was the brainchild of then-Secretary of Agriculture Henry Wallace, and strongly supported by the American Farm Bureau Federation — at least parts of the AFBF.

Targeted primarily at such crops as cotton and grains, the AAA enjoyed the backing of farmers where those crops predominated, mostly the Midwest and South. The Northeast, where dairy products, fruits and vegetables were the rule, saw more harm than benefit from the law. As consumers of feed, for instance, dairymen were profiting handsomely from depressed prices for both grain and cottonseed. They had no reason to cheer any scheme that might raise those prices. How effective the AAA was in raising prices is debatable, but that was its intent.

Meanwhile, the benefits AAA paid farmers to entice them into participating — and to compensate for deferred production — were financed through taxes on processing of the program crops. Those taxes were passed on to the consumer, of course, so for all intents and purposes, Northeastern dairymen were paying taxes to support a program designed to raise prices on products they had to buy.

The issue threatened for a time to dismember the AFBF, and it kept an uncomfortable spotlight on "Triple A" committees, especially in the South. Those committees, made up of local farmers, were responsible for overseeing individual acreage allotments and checking contract compliance, much as Agriculture Stabilization and Conservation Service committees would later monitor conservation and price support programs. The acreage reduction plan and the taxes that supported it also engendered strong opposition among non-farm purchasers of program crops, particularly the powerful textile industry, which naturally had a vested interest in low cotton prices and the same aversion as Yankee dairymen to paying taxes for the undoing of those low prices.

Governors, manufacturers and the textile lobby throughout New England did not like the Triple A program because they felt it was responsible for driving up the price of cotton. The April 22, 1935, edition of *Time* magazine reported: "They went to Washington wailing that unless the cotton processing tax was withdrawn, their industry would be ruined." At that time, AAA administrator Chester Davis told a friend in Washington: 'Well, there's one lobby you haven't heard from. That's the farmer. And when they come, you'll sit up and take notice.'

Davis was right. The lobby against the cotton processing tax provoked a response that united farmers across the country and coincidentally sparked one of the greatest organizing efforts in the AFBF's then-brief history. The response was what came to be known as the "farmers' march on Washington," a May 14 storming of the Capitol by an army of four thousand worried and angry plowmen.

This movement, referred to by the media as "The Great Trek of 1935," resolved itself into a gigantic farmer's lobby that was called by expert, unbiased Washington observers as the largest and most effective of modern times.

The "farmers march" appears to have originated with concerned AAA committeemen, in particular Cliff Day of Plainview, Texas. Longtime Texas Farm Bureau board member Loys Barbour of Iowa Park remembered Day as the individual who planted the seed, and a lengthy treatise on the Farm Bureau during that period tends to substantiate his recollections, at least to a point. The march, says Christiana McFadyen Campbell's

Depression-era farmers, faced with long, rugged roads and poor transportation, appreciated the backing of the government's new "Triple A" program.

"The Farm Bureau and the New Deal," began "among the farmers" but was actually "started by one or two leaders among the AAA production control committees in the South."

According to the May 25, 1935 edition of the *Literary Digest*, "Mr. Day, apparently the leading spirit of the farmer's march, said the march had its inspiration in a 'bunch of fellows sitting around talking, and they collected the money in the schoolhouses and passed the hat.'

"Sun-bronzed, lanky, homespun Cliff Day stood on the Washington scene at the head of an army of farmers 4,000 strong, who massed at the Capitol simply, solely, and hence astoundedly, to praise the Agricultural Adjustment Act."

The farmers gathered on the South Portico of the White House to hear President Roosevelt and other administration agricultural leaders speak, then they crowded into the galleries of the Senate and House. For the first time, farmers realized that, as long as they remained united, they could and would be heard in Washington, even though many went home angry because of the "glorified snubs" they had received from lawmakers.

AFBF, in fact, didn't know quite what to make of the situation. The Farm Bureau's Washington lobbyist considered it "mob action" that could reflect badly on the organization, and he wanted state leaders to discourage their members from participating. AFBF president Edward A. O'Neal, on the other hand, was afraid not to participate, and actually sought a leadership role for the Farm Bureau.

Failure to jump in front of the parade, he warned, would be "turning it over to the other fellow" to "ride to glory."

The "other fellow" in this case was the AAA, which might have been fairly well co-opted in the Midwest where Farm Bureau organization was strong, but was much more independent throughout the South, where AFBF membership was still scanty. For all of its support for the AAA, the Farm Bureau had already begun to worry that the new committees might pose a turf challenge to local farm bureaus, and the large turn out of farmers for the Washington march appeared to bear that out; fully three-fourths of the marchers, a Farm Bureau official estimated, had come from the little-organized South.

One of them, of course, was Cliff Day, another J. Walter Hammond of Tye, near Abilene. They were both AAA committeemen and TAA members, if the then loose-knit group could be said to have had individual members. Barbour recalled Hammond commenting often about the cold shoulder he received in the nation's capital.

"We can't talk to every person who comes to Washington," he was told. "You'll have to go home and get organized." The meaning was clear: without individual dues-paying members, the Texas Agricultural Association could claim to represent no one. If indeed Day was the instigator of the "farmers' march," it had to be a bitter pill to realize that he and those he brought with him would go essentially unheard while others had the government's collective ear.

The Texas delegation took that lesson to heart. In 1936 the Texas Agricultural Association reorganized, firming up the role of county associations but shifting to individual memberships and funding through individual dues. TAA also sought and received affiliation with the American Farm Bureau Federation, something the Texas organization had not had officially since the dissolution of the Texas Farm Bureau Federation three years earlier.

AFBF president O'Neal had previously dismissed Texas as having "too many voluntary farm groups." That made it a poor prospect for affiliation with the national group in his eyes, because as the demise of the old Texas Farm Bureau Federation bore out, farmers who could move freely from one voluntary outfit to another had no strong motivation to stay put and pay dues. O'Neal's position softened, however, in the face of the AAA threat and the organizing rush it spawned. AFBF welcomed Texas back into the fold, just as it actively courted a number of other unorganized southern states.

Attracting and holding members would prove to be a difficult and often thankless chore for TAA leaders over the next several years, and there were to be intramural tugs-of-war as well as assaults from without, but the "modern" Texas Farm Bureau had arrived in fact if not quite yet in name. That, too, would come.

In 1941 the Texas Agricultural Association dusted off and adopted the name "Texas Farm Bureau Federation," using it illegally — if unwittingly so — for the next several years. Somewhere along the line, the group discovered that the "old" Farm Bureau, though dutifully voting itself out of existence in 1933, had failed to notify the Texas secretary of state of its demise. As far as the state was concerned, the earlier organization was still chartered under that name, and the new one had no right to call itself the "Farm Bureau." One of Austin Crownover's first chores as a new field man in 1947 was a bit of detective work. Obtaining the names of the last known directors for the "old" TFBF, Crownover and colleague Marvin Carter fanned out across the state, looking them up.

"About half of them were dead," Crownover recollected nearly a half-century later. "We had to go around to those who were still living in Texas and get them to sign a dissolution so the secretary of state could give us a charter using the name Texas Farm Bureau Federation."

The organization made its final name change in 1954, dropping "Federation" to become simply the Texas Farm Bureau. By then its structure, as well, was firmly set in the Farm Bureau mold: membership was rooted in the county, which in turn was affiliated with the state organization. The state group, likewise, represented its members before the American Farm Bureau Federation. Farm Bureau made considerable mileage out of the claim that its policies and control originated at the grassroots, and that was no idle talk.

TEXAS FARM BUREAU: GROWING PAINS

Texas Farm Bureau membership increased during the latter half of the Thirties, but not at a pace acceptable to either its leaders or members. The organization continued to operate on a shoestring, gaining new members along the way, but losing old ones. The problem, it became apparent to some of the leadership, was that Farm Bureau really had no mechanism nor means to retain members from one year to the next.

As soon as he returned home from the march on Washington, Clifford H. Day, a cotton farmer from Hale County, immediately began to make preparations, including the framing of rules and bylaws, for transforming the Texas Agricultural Association into the current Texas Farm Bureau. Officials at Texas A&M University had noticed Day when his 320-acre farm won first in a statewide contest held to select the best balanced farm in Texas. They named him chairman of the State Cotton Advisory Board and Hale County Cotton Commissioner. When the Triple A was established, Day became a community committeeman in his hometown of Plainview.

In December of 1935, Day was elected vice president of TAA under the presidency of H.G. Lucas and was instrumental in petitioning the American Farm Bureau Federation for membership as the bureau's Texas affiliate. The petition was granted the same year Day became president of the Texas Farm Bureau.

The fight to improve conditions for the farmer was never ending. During this time, the federal government submitted a plan for organizing the Soil Conservation Service in Texas. Day and other agricultural leaders, however, refused to accept the plan because they felt it would give the government too much control. Instead, they framed a soil conservation law of Texas, which made such activities in the state independent of federal control. The law, as Day and other Farm Bureau leaders proposed, was passed by the Texas legislature in 1939.

However, Clifford Day always believed that the government's "Triple A" program had kept the Texas farmer from going under during a dark and economically depressing era. He would later write for *Texas Agriculture*, the official publication of the Texas Farm Bureau:

"The American farmer is the most efficient farmer in the world, in spite of the law and uncertain income he has received so often. In 1933 when the Congress passed the AAA Crop adjustment program, the United States had the greatest surpluses of farm commodities and the lowest prices that were ever known.

"Adjustment meant then exactly what it means now. It means planting and using the necessary acres of farmland to certain crops to assure bountiful supplies without price depressing surpluses...

"In nine years of experience with the adjustment program, farmers have planted and plowed under legumes, terraced, strip-cropped and contoured their land, and as a result, today the American farmland is in a much better physical condition to meet the demands of a World War.

"On top of that, Southern farmers have had nine years experience in diversified farming. Today, many of them are producing beef cattle, sheep, and hogs and chickens, besides a great variety of vegetables and food crops that were never grown on Southern cotton farms before...

The small increase in farm prices and the stabilizing influence of AAA with its commodity loans and crop insurance has made it possible for the farmer to liquidate many of his debts, repair and improve his farm fences and buildings, and put his farm equipment in a fair condition...."

Because of the AAA crop adjustment program, the farmer was beginning to experiment with diversified farming in an effort to improve his land and his economic stability.

In 1938, Day wrote to alternate director Jay Kirkland: "The prospects for the biggest TAA in 1939 are very bright. Congress and the state legislature meet in January and these two bodies are going to do or undo some legislation that will vitally affect the farmers. Now is the time for the farm leadership in Texas to assert itself. If we don't build a militant farm organization, it will be because enough of us have not worked at the job."

A year later, Day's unbridled optimism had not withered. He reported to the Directors of TAA in August: "Now instead of having one field worker, we have twenty-four which will give us the greatest field force we have ever had to build a big strong state organization for Texas.

"I don't believe the field was ever riper and whiter for a large harvest in the Texas Agricultural Association's history. Farmers are gathering a reasonable crop in most sections and the AAA is now ready to distribute 1939 parity checks on cotton and have already distributed the parity checks on wheat.

"We feel like we have a just claim in asking farmers to support an organization that is directly responsible for bringing this extra money to Texas farmers in the way of parity payments for 1939, to say nothing of already providing for parity payments in 1940.

"To this date we have 11,331 members in the state for 1939. With as many workers as is now provided for, we should encounter no trouble in securing enough members to make a total of 20,000 1939 members by December 1. Your whole-hearted cooperation and active support are prayerfully solicited, in reaching this goal."

During the late 1930s, Day was at least as much a "Triple A" man as he was a Farm Bureau man, and AAA's relationship with the Farm Bureau movement was beginning to sour.

It was a gradual process, but an inexorable one; as late as 1936, AFBF president Ed O'Neal was described as "panicked" over a crucial U.S. Supreme Court ruling that overturned the heart of the Agricultural Adjustment Act of 1933, under which AAA operated. Farm Bureau was a major player in a stopgap replacement program which attempted to circumvent the constitutional issue by shifting the rationale for acreage controls production limitations to conservation. O'Neal and Farm Bureau were also prime movers two years later when Congress passed a more permanent replacement, the Agricultural Adjustment Act of 1938.

In both instances, however, Farm Bureau's interest was in the fate of the program, not the bureaucracy administering it. In fact, as O'Neal and other organization leaders had recognized at least as early as 1935, AAA and its burgeoning sister agencies at USDA were beginning to challenge Farm Bureau for dominance in agricultural policy. And it was not just a battle for turf, it was a battle for philosophy: despite regional stresses and strains, national Farm Bureau policy actively supported "emergency" New Deal programs of government intervention and even production controls, but it did not intend the most intrusive of these to outlive the Depression.

Meanwhile, Farm Bureau was adamant that agriculture policy be determined at the grassroots — county Farm Bureaus would pass their members' wishes on to the state federations, and those in turn would carry the word to the American Farm Bureau Federation. AAA and the other New Deal agencies were predicated, by contrast, on the top-down, central planning notion that Washington and its "professional" staffs knew what was best for agriculture, and they clearly had no intention of fading gracefully away when the emergency was over.

Little documentation survives from the Texas Agricultural Association during the 1936-39 era, and recollections of old hands who were on board at the time are sketchy, but the record indicates that Day and the TAA's board were on diverging paths; just how far those paths eventually diverged will be explored at some length later. Suffice it to say that by the time of the TAA board meeting on Sept. 23, 1939, the organization and its president had come to a parting of the ways.

Twenty-four days after writing the board about his optimism for securing a membership of 20,000 by year's end, Day submitted his resignation letter to the directors and alternates of TAA, writing: "I have always said that a strong militant farm organization was one of the most essential things in the solution of the farmer's problems.

"Today, I am more convinced than ever that this is true. The principles and purposes of the Texas Agricultural Association and A.F.B.F., to

A band of Texas Farm Bureau field men—including Leon Lane, O.K. Hoyle, J.D. Jordan and Melvin Dansby—involved themselves in a major membership drive as TFB made strides to become the nation's largest state farm organization.

me, are the nearest approach in that direction.

"I think I have tried to give the best I had in helping build such an organization. That I have made mistakes, no one will recognize quicker than myself.

"To reach the goal we have set for Texas has been hard, it will continue to be hard. I am sure you fully recognize this fact from your own personal experience.

"I still believe the job can be done and done well. The determined efforts of a militant leadership can put the job over.

"The purpose of this letter is to say that I find it impossible to care for my family on the funds provided for me inasmuch as I have not collected one penny to date.

"I am seeking employment that will give me some cash income.

"I trust that whatever my new field of endeavor will be, that it will afford me the opportunity to contact farm people. If I possess any good for future agriculture, I still want to give my mite."

By 1943, Day had become the senior administrative assistant for the Agricultural Adjustment Agency.

TAA vice president Romeo Korth declined to succeed Day, and the board chose Walter Hammond from among its own ranks. He would hold the office of president for almost two decades, longer than any other individual. His was not an auspicious beginning, however. Loys Barbour recalled that Hammond's wife chided him for accepting the leadership of an organization that had only five thousand members, was $5000 in the red, and had "no potential at all."

As was true of many of his contemporaries, Walter Hammond had little formal education. He was raised at Tye, near Abilene, on the eastern edge of West Texas, which was still something of a frontier region in his childhood. Hammond was already a teenager when the new century rolled in, and there simply wasn't much formal education to be had at that place and time. There was a wealth of informal education, however, and Hammond was a man who took advantage of all there was to learn. As a young man he adopted "advanced" farming techniques such as terracing, and he persisted in employing them despite "expert" advice to the contrary. His confidence, more often than not, was rewarded with success.

Self-motivated and staunchly independent, Hammond had a persuasive nature that was to serve the Farm Bureau well for years to come. Unlike his predecessors, his approach to the post of TAA president was to focus on the individual producer, not on government programs. Hammond was an able proponent of the organization's various cooperative marketing efforts, and he had strong opinions about how government should treat agriculture, but his inclination was always to place the farmer firmly in charge of the program rather than the other way around. That attitude revolutionized the organization's philosophy, setting it on the course it follows to this day.

At first, Hammond's role was somewhat limited; he toured the state, selling the idea of agricultural organization in general and TAA in particular, while the organization's nuts and bolts work was conducted by a succession of "general managers" and "executive secretaries." None of them was quite satisfactory to Hammond, however, and he finally took full charge of the reins during the group's January 1942 convention, a year after the Texas Agricultural Association became the Texas Farm Bureau Federation. At Hammond's urging, TFBF directors addressed the group's perennial financial pinch by more than doubling membership dues, from two dollars to five dollars per family.

That alone didn't do the trick. In fact, it drove off so many members that Farm Bureau barely held even in net dollar income despite the dues hike. The organization at the time had a ready-made issue that should have rallied prospective members — protecting agricultural prices from being crushed under the government's wartime price ceilings. Instead, memberships plunged. TFBF lost more than four thousand members in 1942 after the dues increase went into effect, leaving rolls at barely 11,098 families. In 1943 the level dipped below 8,533, and at its low point in 1944, membership declined to just over 7,884. Human nature, it appeared, was at work.

The price ceilings battle, along with fights over parity and price supports, were clearly important issues to many farmers, but few could see a direct cause-and-effect relationship between writing a check for dues and winning a skirmish in Washington. Farm Bureau was fighting the battle with or without them, it was easy to rationalize, and that five dollars could come in handy at home.

No one liked to be singled out as a slacker, of course, so membership rallies with their peer pressure would periodically convert sizeable numbers of reluctant joiners. Farm Bureau couldn't maintain those memberships, however, without some means to keep farmers coming back year after year. Walter Hammond found that means — actually, the first of many — in 1944 in the form of health insurance from Blue Cross-Blue Shield.

Health insurance was relatively uncommon prior to the war years, but wartime wage controls left industrial employers in particular facing huge demands for not only material, but for workers, the labor pool sorely depleted by the armed forces' own massive manpower needs. When he couldn't offer higher wages than the next fellow, the desperate employer was forced to sweeten the bid with fringe benefits.

One of the most successful of the few allowable fringes turned out to be health care. The health insurance industry mushroomed, and demand quickly spread beyond the industrial sector. Then as now, however, individual policies were far more expensive than those purchased at group rates. That left the independent farm or ranch family at a distinct disadvantage; even the largest agricultural employer couldn't deliver enough premiums to wrangle much of a rate break.

It was Walter Hammond's common-sense, country boy genius to

recognize the potential of using Farm Bureau's membership to command favorable group insurance rates. He then used the insurance program to attract — and hold — new members. Initiated in April of 1944 through Dallas insurance agent E. R. McBee, the Blue Cross-Blue Shield hospitalization plan cost each participating Farm Bureau family $5.25 per quarter and covered children as well as parents. Non-members, of course, were not eligible, and once a member had paid a few quarterly premiums, his investment in the policy gave him a strong incentive to "re-up" when dues time rolled around again.

Texas Farm Bureau sponsored membership rallies, conventions and dinners on the ground in order to get the TFB message to as many farmers and ranchers as possible in the state.

Armed with the new tool, Hammond hired Texas A&M graduate Marvin Carter as TFBF's first "Director of Organization." Carter was employed at the time as the Hartley County Extension agent, and had previously helped organize for the Farmers Union in East Texas. Carter served as Hammond's right hand and organizing guru for a decade before joining the foreign agricultural service for a stint in the Middle East. His Farm Bureau career was almost cut short, however, by tragedy.

Less than two weeks before Christmas in 1946, with the United States still celebrating the end of World War II little more than a year before, the Texas and Louisiana Farm Bureaus had chartered a special train to carry members and officials to and from the American Farm Bureau Federation convention in San Francisco. On the return trip, the train was approaching Salt Lake City just after dawn on Saturday, Dec. 14. Most of the conventioneers were still in their berths. The day's agenda called for a layover and a sightseeing excursion in Utah's capital. It did not call for them to share the track with an outbound freight train.

Investigators later determined that the freight had been scheduled to switch off at a siding but was running behind; as he neared the switch, the freight's engineer sighted the oncoming special and realized he could not pull his own string of fourteen slow-moving cars onto the siding before the passenger train arrived. Instead, he brought the entire train to a halt and began the ponderous process of backing up, hoping to buy time until the passenger train could stop. The other engineer sighted the freight as well, and began slowing down. It takes a lot of trackage, however, to slow hundreds of tons of hurtling train or to accelerate many more hundreds of tons in the other direction from a dead stop. On that crisp morning outside of Salt Lake City, there just wasn't enough track between them.

The two trains collided head-on. The speed differential was down to a sedate twelve miles per hour, but the passenger train's weight and momentum converted that little velocity into tremendous energy, telescoping the lead sleeper into a baggage car and derailing the two, and throwing passengers in the other cars around like dolls. The wreck sent seventeen people, mostly Texas Farm Bureau members, to Salt Lake City hospitals for treatment of significant injuries, and scores of other passengers were treated on the scene for cuts and bruises.

Marvin Carter was in the most seriously damaged car at the time of the crash, but escaped with only minor injuries. Like a true organizer, Carter was credited with arranging the orderly evacuation of passengers throughout the stricken train. Other heroes included Mrs. Loys Barbour and a Louisiana Farm Bureau wife, both trained nurses, who ministered to the injured without even stopping to dress.

In some ways, however, the entire Farm Bureau contingent qualified as heroes. The *Salt Lake Telegram* remarked with wonder at the lack of panic among passengers, then quoted Mrs. Clarence Simpson of Coleman with the explanation: "We're from Texas," Mrs. Simpson said. "We don't get excited. We take things as they come and make the best of it." Mrs. Simpson made her comments at the hospital while awaiting news of her husband, who had been trapped for some time in the wreckage of the crushed Pullman and was being treated for then-undetermined internal injuries.

In the months to come, Farm Bureau also "made the best" of things. With Carter in control of organizing, the group soon began adding field men to its staff, gambling with money it didn't have that the new employees could enlarge the membership enough to at least pay their own salaries. It took time, but the gamble paid off.

The first member of the new field staff to "stick" was Austin Crownover, a schoolmate of Carter's who, like Carter, had gone into Extension work. His duties involved local administration of the AAA programs, and when those were split off from Extension, Crownover went with them. He was in Hale County when Farm Bureau backers there launched a membership drive to boost their meager numbers, coordinating it with a statewide effort that even had Hammond broadcasting over Fort Worth radio station WBAP.

Crownover's Extension experience predisposed him toward TFBF, so he volunteered to assist. The Hale County drive was representative of

County leaders in small communities throughout Texas formed the backbone of the organization.

many during those days when organizing county Farm Bureaus was akin to pulling hen's teeth. With Crownover's help, organizers had amassed a long list of producers and invited them to a meeting in Plainview to hear Hammond's "kickoff" address on radio.

"Only about six showed up," Crownover recalled decades later, and most of that six "sat around and said they didn't guess farmers could organize. They wouldn't stick together, they just weren't interested. They poured out all the pessimism that I ever heard in my life. It kinda' ticked me off."

Crownover— who, as a volunteer, had no stake in the matter — responded with a pep talk. He chastised the naysayers and countered that they could organize if they'd just make up their minds they wanted to.

Local farmer John I. Boling called Crownover "the craziest damn government man I ever heard talk."

But then, he agreed with everything that Crownover had said. Boling's conversion broke the ice, and the group voted to reconvene at Hale Center, where more farmers were expected to turn out. But the head count there was still just a dozen. Nevertheless, that hardcore few scheduled a drive for several weeks later. Meanwhile, Crownover, joined by the local county

In order to increase membership, Texas Farm Bureau even turned one drive into a baseball theme, as exemplified by Jerry Bedenbender, Ed Cumbie, A. C. Machemehl, Jr., and Cliff Laywell.

agent and another volunteer, conducted thirteen community meetings throughout the county to drum up support and sign helpers for the door-to-door push. The allotted day dawned with a blizzard that closed roads throughout the area and virtually shut down travel. Undaunted Crownover and his crew — still volunteers, all — took their message to the airwaves, broadcasting their Farm Bureau pitch over a Plainview radio station.

They announced a one-week postponement of the physical drive itself, and when it finally came off, Crownover had amassed almost eight hundred new Farm Bureau members.

"This had never happened in Texas in the Farm Bureau," Crownover said, "and I didn't know it until years later, but the board of directors told Walter Hammond to go hire me, it didn't make a damn what it cost."

At least it didn't matter to the board. Walter Hammond, however, still had his eye firmly on the bottom line, and his "negotiations" with Crownover offer considerable insight into the character of the man who nursed Texas Farm Bureau through its growing pains.

"Walter Hammond was a horse trader," Crownover said. "He was the best in the world." He was also "a miser on money" who "saved every penny he could, including my salary." The two men and their wives met at Post, about halfway between Plainview and Hammond's home near Abilene. From about eleven o'clock that morning until sundown, Hammond talked Farm Bureau and organizing.

"I bet if he said it once, he said it a dozen times," Crownover recalled: "Now, we would have enough money to pay your salary and Marvin Carter's salary for one year, if we don't get any more memberships. We've got enough money that we can pay you for that long."

But, Crownover added, Hammond also made it a point to poor-mouth Farm Bureau's finances.

"He kept talking about how we didn't have much money but we could grow, and all this. He kept talking and talking — never did make me an offer. What he was doing was selling me so I wouldn't ask him too much. He had been ordered by the board to hire me, but I didn't know it."

As the hour grew late, Crownover consented to the job and — in the interest of getting both parties home at a reasonable hour — offered to call Hammond back later with a salary proposal. That was just what Hammond had wanted, for Crownover to sleep on the matter over a weekend, meditate a bit on Farm Bureau's modest but promising financial condition, and talk himself out of any overly ambitious salary notions.

It worked, and Crownover ended up asking for a considerably slimmer salary than he had originally intended to seek, all without the shrewd Hammond ever confronting the issue directly.

The Texas Farm Bureau Federation staff at that point consisted of Carter, Crownover, and Hammond's secretary, Creola McGill.

Across rural Texas in the 1940s, farmers and ranchers were facing a growing, critical need for insurance services. Only a few scattered companies

Texas Farm Bureau knew it had to provide insurance to those farmers whose crops—like this wheat field—grew in remote sections of the country.

were even paying any attention to those who had settled along the state's backroads, growing the crops and raising the beef that kept food on the nation's tables. Most agents preferred doing business in cities and throughout the suburbs where populations were concentrated. Almost none of them had any interest in driving down lonely, isolated roads to a remote farm or ranch in the hopes of perhaps writing a single policy. It just did not make good business sense, or so the large insurance companies seemed to think. Besides, history had showed rural areas did not have the benefit of fire protection and police protection, and, as a result, generally had a higher incidence of total loss in case calamity struck. They were a risk that most insurance companies simply wanted to avoid at all costs.

So farmers and ranchers were searching for better insurance.

And Texas Farm Bureau was committed to providing a better service to its members, even if that included insurance, and, in time, it certainly did.

As the health insurance program's drawing and holding power became apparent, Farm Bureau branched out to other member services. Along with several other state Farm Bureaus, the Texas group chartered a life insurance company in 1946. The Southern Farm Bureau Life Insurance Company listed Texas, Alabama, Arkansas, Kentucky and Mississippi as participants. With headquarters in Jackson, Miss., the company was headed by Dave Mieher, former director of sales for the Illinois Farm Bureau's insurance programs. When Mieher retired more than two decades later, the insurance company he helped form had more than two billion dollars worth of policies in force. As soon as the life company was on solid footing, Mieher also helped organize the Southern Farm Bureau Casualty Insurance Company to protect members' investments in vehicles and other property.

Early insurance sales were underwhelming, however. As Farm Bureau began to hire organizational field men, it licensed them as insurance agents as well, and sent them out to do double duty, selling both the Farm Bureau and its insurance products. The tactic, attributed to Loys Barbour, was born of necessity because the insurance programs' very newness made them suspect among professional insurance agents. With no history and a prerequisite of membership that served to limit the pool of potential customers, Farm Bureau policies did not inspire enough confidence among salesmen to encourage them to abandon established lines and throw in with TFBF.

For all its apparent practicality, however, Barbour's idea of sending field men out to sell insurance was an attempt to make orange juice out of apples. Many good membership organizers simply weren't good insurance salesmen.

"I remember," Crownover told an interviewer, "when we had some seventy agents" and were writing "maybe a $5000, maybe a $10,000 policy" weekly, statewide. "We weren't getting very far very fast."

The situation hadn't escaped Mieher's worried attention, and at his urging, TFBF split its forces. By 1949 it had grown its own crop of insurance agents — if a small crop — and men with that inclination were shifted to full-time sales work. Others, whose tendencies lay more toward organizing, became full-time field men. Marvin Carter's calling was organization, so, again at Mieher's urging, Carter abandoned the sales efforts which had grown to monopolize so much of his time, and returned to what he knew best. Crownover, who had good sales skills but had gradually picked up Carter's organizing load, unlimbered that burden and became Director of Sales for Texas Farm Bureau insurance companies.

Farm Bureau's list of services expanded yet again in 1950, when the board of directors voted to raise capital stock for a fire insurance company. The initial plan put forward by the staff called for a mutual fire insurance program, but board members balked at the potential exposure such an arrangement implied, and elected to go the stock route. To assure control of what became chartered as the Texas Farm Bureau Insurance Company, the organization retained all common stock and sold only nonvoting preferred shares.

Meanwhile, Farm Bureau's insurance sales force had come into its own. By 1950, Crownover had four district sales managers and a growing list of insurance agents. That February, he convened the group's first agents' meeting in Abilene to teach the new staff how to rate and underwrite Farm Bureau Insurance's brand of policy.

Dee Knox was Crownover's first district sales manager. A field man when sales and organizing were still combined, Knox had been a key factor in Crownover's decision to accept the sales director's post. As Crownover told the story, when the field staff split was initiated and the sales director post offered, he invited Knox out for a hamburger and to discuss the proposition. One provision was that Crownover and Carter, in switching responsibilities and splitting the field force, could take the field men they wanted into their respective divisions. Crownover wanted Knox, who he termed "the only agent who had ever written anything in Texas that amounted to anything."

"I told Dee what was going to take place," he recalled decades later, "and asked him ... 'do you want to go with me or go with Marvin Carter?' He didn't bat an eye. He said, 'I want to go with you.'

"I said, 'Dee, that made up my mind. I want to be Director of Sales.'"

With the new management structure in place and a growing network of agents writing policies, Farm Bureau's insurance program shook off the doldrums. The life company had just over $3 million worth of policies in effect in 1947, but by 1949 the total was up to $5.6 million. The next year's tally on life policies was more than $8 million, and the casualty company had 18,310 vehicles under casualty coverage, along with 365 policies in force in the fledgling fire program. Within five more years, life policies were worth nearly $40 million, the casualty company covered sixty-seven thousand vehicles, and fire policies numbered more than eighteen thousand.

By the late 1950s, Farm Bureau board members began to rethink the structure of the fire insurance company. As successful as it had become as a stock company, it nevertheless did not offer the level of service originally anticipated. Bowing to the wisdom of the staff's original proposition, they converted the Farm Bureau Insurance Company to a mutual company effective Jan. 1, 1959. They still limited Farm Bureau's potential exposure, however, by chartering it as a "legal reserve non-assessable" mutual.

That spring, the revamped Texas Farm Bureau Mutual Insurance Company expanded beyond fire coverage to add crop hail protection, and soon became the largest crop hail insurer in Texas. Within two years, policyholders had earned a ten percent dividend. It was the first time in the history of crop hail coverage in Texas that a company paid a dividend.

The company's success forced better performance by the competition, and in short order the writing of hail policies became a cut-throat business. Before long, TFB Mutual reached the point where there was no longer any fat to trim; the only way it could continue reducing rates and improving coverage as the market demanded was to reduce the regulatory burden. In 1964, TFB organized the Texas Farm Bureau Management Corporation and Texas Farm Bureau Underwriters, exempt from regulation by the Texas State Board of Insurance. Over the ensuing years, TFB Mutual added policies covering inland marine, poultry and other risks. In 1992, for "economic and regulatory reasons," the company closed out its crop hail policies, and began writing this coverage through an outside brokerage firm.

Farm Bureau's insurance business has continued to grow through the decades. In 1970, $473 million dollars of life insurance policies were in force in Texas. The casualty company covered 169,141 vehicles, and nearly 58,500 fire policies were in force. The next two decades showed tremendous growth. By 1990, life policies numbered 143,918, and protected Texas Farm Bureau members with over $54 billion of life insurance. The casualty company was covering 475,425 vehicles, and 173,250 fire policies, valued at more than $74 million.

As Walter Hammond and the TFB board had anticipated, memberships increased dramatically because of the insurance program. In 1944, the first year TFBF offered health insurance, Wilson County became the first in the state to establish a county Farm Bureau office staffed with a full-time secretary. Statewide memberships at that point totaled barely more than 7,800, and new-hire Marvin Carter's salary was contingent upon raising membership enough to generate the dues to pay it. Two years later, membership had more than doubled to 18,653, and by 1947, it was up to 27,821.

Austin Crownover's experience with the Hale County membership drive was evidence that the increase in members didn't come easily. Dee Knox told of a membership drive in Guadalupe County. It was in a neighboring district, not his own, but fellow field man Leon Lane had invited

Knox to speak at the kick-off dinner. Knox estimated that Farm Bureau fed ninety people that evening, all of them supposedly foot soldiers in the next week's daily door-to-door drive. Both men were proud of the turnout and already counting the scores of new members Lane would add to his rolls with such an army of volunteers to knock on doors and sell Farm Bureau. When Knox called back after the drive, however, a much more subdued Lane told him that the weeklong effort had netted just one new member.

Wayne Little joined TFB as a field man-insurance agent in 1949 and eventually became Director of Organization. In his early days, he told an interviewer forty years later, Farm Bureau didn't have enough field men to "localize any given area," so the field men it did have covered the entire state.

"We'd start in the lower part of the Valley," Little recalled, "then we would move north as the season moved. In the fall of the year we would end up in the Panhandle. It was not an unusual thing in those days when we went out to work on the Plains we'd go out there and spend two weeks before we'd come home."

With its wide-open spaces and relatively sparse population, West

Texas Farm Bureau supplemented its insurance for vehicles with a strong, statewide safety program. Left to right, representing Texas Farm Bureau, are Joe Smetana, B.C. Jones and Paul Edwards.

Texas presented special logistical problems. "I would set up a meeting at ten o'clock in the morning, one at three in the afternoon and one at seven that night, in three different counties. And then when we finished with that meeting at night — it would be maybe nine, nine-thirty or ten — I'd drive to the next county so I wouldn't have to get up early."

Ed Cumbie of Bronte joined TFB as a field man in the mid-1950s to supplement the drouth-shriveled income from his stockfarming operation. The state had begun to be broken out into districts by then, and Cumbie was given the sprawling West Texas region, stretching from Brown and Eastland counties all the way to El Paso. He recalled a get-acquainted session during which he and field men from other states compared their territories and discussed organizing efforts.

"The fellows from Illinois, Iowa and some of those states up there only had one or two counties," Cumbie recalled. "Of course, they might have fifteen thousand members in those one or two counties." Mouths dropped open when Cumbie told them with all honesty that his district was 450 miles long, 300 miles wide, and contained twenty-six organized counties.

"Golly," said one field man. "That's bigger than our whole state."

And those long stretches of road could be short on amenities. Little told of the excitement among field agents at the news that a certain motel in Georgetown was planning to install air conditioning, and remarked upon the scarcity of good restaurants throughout the state in the late forties and early fifties.

An old tractor on a lonely farm symbolized the wide-open spaces in Texas which presented special logistical problems for Farm Bureau.

"I read an article after I started driving statewide," he said. "It must have been about 1952, and it named three decent restaurants in South Texas. They said everything else was a 'greasy spoon'."

That was a time when TFB board members might be found setting up chairs at the annual convention, so the field men never thought of putting on airs. "Our motto was: We do whatever has to be done," Little explained. And what had to be done the most, for the field men, was selling Farm Bureau memberships.

Little did the job with the help of one of Walter Hammond's anecdotes about watering horses. As Hammond told the story, he sent a man to water one of his horses, and the man came back with a still-dry horse. "Hammond," he said, "this horse wouldn't drink. I led him down there." Hammond's response was, "Your job was not to make him drink. Your job was to make him thirsty, and he'd drink on his own."

"That's just about the way we were," Little said of the early TFB field men. "Our job was to make them thirsty."

And they did. Texas Farm Bureau membership had increased to just shy of 52,978 by 1950, and was more than twice that a decade later. The head count in 1970 was better than 121,000, and it hit the 214,197 mark in 1977. Just four years later, in 1982, Texas Farm Bureau became the first in the nation to reach 300,000 members. Don Neumann, field services director at the time, recalls, "Achieving that 300,000 membership mark had been a primary goal of Farm Bureau for about three years. The success of our effort can be attributed to the hard work of our county leaders who made the goal a total team effort, including the insurance agency force and the county office secretaries. Many of them went out and signed up a hundred new members each, which showed a lot of personal dedication from our people all over the state. Ninety percent of the battle is always attitude, and everybody had the right attitude to determine what job needed to be done, then do it.

"There had always been a lot of pride among the Texas Farm Bureau leadership to become the largest state Farm Bureau, and we had accomplished that. When we broke 300,000, Texas Farm Bureau became a larger player at the national level, and people were more responsive to the rest of our programs."

The decade of the nineties dawned with Farm Bureau's rolls at 320,272. By way of perspective, the number of farms and ranches in the state was declining dramatically even as TFB's membership increased. From 345,000 operations in 1950, the farm and ranch inventory fell to 212,000 by 1970 and to 186,000 by 1990.

Texas Farm Bureau: Getting the Word Out

From the beginning, Texas Farm Bureau has recognized the importance of getting its message across — both to its members and to the public at large. The Texas Agriculture Association. established the twice-monthly *Farmers' Banner*, picking up where the "old" Farm Bureau's *Texas Cooperative News* and the even earlier *Texas Farm Bureau News* had left off.

As is the case with other records, however, few early *Banners* survive, in part because the TAA had no permanent home in its formative years. Moved from the old Farm Bureau's Dallas base to TAA's first president H.G. Lucas' Brownwood home, the Texas Agricultural Association's records, such as they were, later traveled in Walter Hammond's car trunk; wherever Hammond happened to be was TAA's de facto headquarters, and that type of arrangement does not make for the accumulation of extensive archives. Furthermore, Hammond wasn't given to salting away material for posterity. On more than one occasion he was heard to remark that he was less interested in what happened yesterday than in what was going to happen tomorrow.

In fact, the *Banner* itself apparently gave TAA a "headquarters" when it otherwise had none. The earliest editions were printed in Brownwood, where TAA was centered under Lucas. At some point in the late 1930s, however, a Waco printer took over publication, and that print shop appeared as TAA's "official" address before the association moved into the Waco Chamber of Commerce building in 1940.

It was also in 1940 that the *Banner* changed its name, adopting the *Texas Agriculture* moniker that it wears today. The Aug. 1, 1940 issue, number 18 of volume five, announced the name change along with a reduction in frequency to once a month. Quoting TAA general manager W.M. "Bill" Deck, the final issue of the *Banner* announced that the change "came at the request of several thousand members who think the new name will be much more suitable." Deck said the shift to a monthly rather than twice-monthly frequency was intended, in part, to keep TAA in step with other AFBF state affiliates, "nearly all" of them published only once a month. "By issuing a paper only once a month," he added, "a much better paper can be published at a more economical rate."

That latter consideration was of no small concern to the financially strapped TAA, and the "Farmers' Forum" column inside the paper went into considerably greater detail in explaining the need to trim publication costs. The change, the "Forum" revealed, "has been debated, cussed and discussed for the past two years," during which time "everyone seemed to be in favor of such a change as this, but nobody seemed to do anything about it, as Mark Twain once remarked of the weather."

Under the twice-monthly schedule, the column continued, "after the printing and mailing ... had been paid each month, the budget allotment ... was practically used up, and there was no money for other things that make an organization paper what it should be." Of particular concern, it appeared, was the dearth of photos. Today's offset printing processes handle artwork and photography as easily as they do type, but the presses of the Thirties and Forties required a much more complex and costly process of engraving photographic images on metal plates.

"We have received several pictures from time to time," the article said, "of various TAA activities throughout the state that we couldn't run in the *Banner* because our printing expense would not permit." In addition to pictures, "there have been meetings that we would like to have covered personally for our readers, but here again our budget wouldn't stand the strain."

The "Forum" promised "a more personal touch" under the new

An old-fashioned linotype operator sets type for Farm Bureau's house organ, a publication which said "all farmers may not agree with all the statements made, but at least they will hurry to read it each time."

arrangement, in addition to more pictures. "We will do our best," it added, "toward getting around to some of our state leaders and 'get them out on a limb' on state and national questions ... for the edification of our readers ... Let's have a publication," it said, "wherein all farmers may not agree with all the statements made, but at least they will hurry to read it each time!"

The old Farm Bureau's house organ had worn its policies on its sleeve, so to speak, trumpeting in the masthead in 1921: "We Ask A Fair Price For Farm Products Based On The Production Cost Plus A Profit." By 1923, verbosity was run amok, and the masthead declaration had swollen to: "The Orderly Merchandising Of Our Crops As A Demand Is Created Is The Only Way For Us To Get For Ourselves Enough Money To Pay Our Debts, Buy Good Farms, Build Comfortable Homes, Give Our Families The Comfort Other Men's Families Have, Educate Our Children To Become Useful Citizens And Bring Them Up As God-Fearing Men And Women."

That somewhat windy statement epitomized the old group's cooperative marketing focus for all who might wonder, but it apparently also soured the leaders of both the "old" and "new" organizations on the idea of verbose mastheads; after a run of several years, the motto was dropped, and it was at least a dozen more years before another such statement — this one much briefer and more tasteful — appeared on a Farm Bureau publication. Beginning about 1942 and running until August of 1959, the *Texas Agriculture* masthead featured, in banner form, the motto: "Full Parity For All Farm Products."

During the 1940s and 1950s, President Walter Hammond, left, used the motto: "Full Parity For All Farm Products."

The primary responsibility of an association publication is to keep members informed of developments within the organization — meetings, personnel changes, election results, policy statements and the like — and *Texas Agriculture* fulfilled that obligation. A typical issue from the late 1950s notes the selection of a new president for the Williamson County Farm Bureau, a membership drive in Hopkins County, and even an "anti-rat campaign" in Erath County.

But the publication didn't stop with routine internal housekeeping. The same 1950s issue contains an article by TFB Research Director W.C. "Bill" Wedemeyer explaining "The Truth About the five Billion

Dollar Agricultural Budget." Wedemeyer noted that the national press had been having something of a field day with claims that "the government is spending about five billion dollars a year to help farmers." In reality, he told *Texas Agriculture* readers, "much of the five billion dollars is spent for groups other than farmers."

Wedemeyer noted that $100 million went to the school lunch program, $64 million went to food programs for "needy people," and $234 million was spent on food aid to foreign countries. Another $637 million was included in the ag budget for related foreign aid, he wrote, $120 million for the Forest Service, and hundreds of millions of dollars actually were earmarked to guarantee agriculture-related lending programs for which the government would eventually be reimbursed with interest.

In addition, *Texas Agriculture* included such regular columns as Walter Hammond's "Blowing Off The Foam," and Marvin Carter's "Covering Texas With Carter." Carter, writing during the Forties, reported on organizing activities and shared tidbits picked up in his statewide travels. Despite the fact that he was preaching to the choir, Carter seldom overlooked an opportunity to plug the value of Farm Bureau membership, even if he had to reach a ways to make the connection. In the April 1949 issue, he gave a "hat's off" to "the Neiderwald Community" of Caldwell County. "If I've ever seen a rural preacher ... putting his heart and physical efforts into community development," Carter enthused, "it is Rev. Kloppe. As often said, 'next to the church, Farm Bureau has the most to offer farmers.'"

Hammond's columns often outlined his positions on issues of the day, whether agriculture-related or not. The May 1949 edition addressed the legislature's progress on the farm-to-market road bill and exhorted Farm Bureau members to punch up their state representatives in support of its passage. It also commented unfavorably on such non-farm matters as a squabble between the military branches over what would later become consolidation under the Joint Chiefs of Staff, and a threatened auto workers' strike. The latter had Hammond quite concerned and more than a little upset.

New car production had been suspended during World War II, and pent-up demand kept manufacturers back-ordered long after the war was over. That situation was working itself out by 1949, but Hammond observed that "Just when folks begin to be hopeful that maybe an automobile might be bought from the regular dealers without dickering with new and used car lot dealers, who somehow manage to get possession of most of the new cars and then up the price, workers threaten to strike!"

Farmers "would be denounced," he wrote, if they were to do the same thing.

More important than its role in simply spreading Hammond's personal view of the world, his "Foam" column served to shape the outlook of the membership, and to support and encourage like-minded members.

In May of 1951, Hammond sank his teeth into a government controversy that would repeat itself — with disastrous results to the beef industry — two decades later, and which another two decades along seems almost inconceivable in its wrongheadedness.

The issue was wage and price controls, prompted by post-war inflation and spurred by the communist threat then manifesting itself in the Korean conflict. The nation was still in the process of standing down after World War II, and the incredible feats of material and manpower management accomplished during that global struggle gave many Americans an exaggerated opinion of their government's capabilities; what it had managed to do for a brief time under the influence of intense patriotic fervor, they thought, it should be able to accomplish routinely without benefit of such powerful inducements.

Hammond didn't quite see it that way. In his "Foam" column, he noted with displeasure that in the then-current "emergency" it was "generally conceded that production is the most important factor in avoiding all-out war and retarding inflation." Consequently, he continued, the Truman administration had created a plethora of "boards, committees and agencies" to "ferret out the most essential products," encourage their production and discourage the production of non-essentials. Clearly, the old wartime command psychology was still in full plumage.

Hammond went on to chide the administration for "lacking the courage" to deal with inflation in more imaginative ways, rather than plunging ahead with outright "price fixing."

"In the first place," Hammond wrote, "price fixing does not remove the cause of inflation," and "in the second place, the usual run of price fixers have no practical experience or knowledge" of the myriad of interlocking factors "in the production of essentials which must of necessity be produced in various and widespread areas of the nation."

He noted that the Truman price fixers were quick to "allow increases in the wages of labor and prices of materials," but then denied producers the opportunity of passing those increased costs along. One clear example was an apparent willingness to capitulate to meatpackers' union demands for higher wages while live cattle prices were to be rolled back three separate times.

"The price fixers are all mixed up," Hammond contended. "They seem to realize that you must pay people for processing, but do not seem to know that if you do not pay people to produce, there will not be anything to process."

And, he added, "If you think that is a bit screwy, listen to this: When increased wages of industrial labor draws domestic labor from the ranch and farm, it has been the practice of farmers and ranchers to resort to untrained Mexican nationals... to supply the needs. Now, legislation making it a crime for farmers and ranchmen to employ Mexican 'wetback' labor has been recommended to President Truman by his commission on migratory labor. Suggested penalties ... include fines and imprisonment."

In essence, the government was charging boldly forth with a multipronged program that would squeeze agricultural producers from all sides and make it almost impossible for them to produce food and fiber, all the while exhorting them to step up production.

"This is simply giving you the facts," Hammond concluded with a touch of mischief, "and anything in this column that would indicate stupidity in government is purely incidental."

Texas Agriculture and its predecessor publications were compiled and edited in the early years by a succession of Farm Bureau administrators. As the organization began offering services and finally turned the corner on membership, however, it grew too large and complex to be run by the skeleton staff of the Thirties and Forties. Various functions swelled to require their own departments, and in 1952 the Texas Farm Bureau Information Department came into being under the direction of Bill Hoover.

Hoover came to the Texas Farm Bureau after a short stint with the Memphis, Texas *Democrat* newspaper. He provided the media expertise necessary for the development of two publications, *Texas Agriculture* and *Texas Neighbors*, a state-of-the-art radio-TV department and a newsletter service for county Farm Bureaus.

The Information Department at first included only print media, but that situation didn't last long. On May 24, 1954, the first "Farm Bureau Roundup" radio broadcast, originating from a garage, went out over the airwaves on sixty-two stations. Built around on-the-spot reports by a team of agricultural reporters from across the state, the weekly broadcast in time

Texas Farm Bureau was using radio microphones all over Texas in order to get necessary information out to its members and prospective members. A special broadcast over WBAP in Fort Worth signalled Farm Bureau Week, featuring talks by (left to right) Frank Taylor, Farm Bureau vice president; Walter Hammond, Farm Bureau president; P.T. Price, a director; Layne Beaty, WBAP announcer; Raymond McElrath, a director, and Loys Barbour, a director.

came to be carried on more than one hundred radio stations.

Under Hoover's leadership and guidance, these regional, on-the-scene reports chronicled the week's happenings in Texas agriculture from the Rio Grande Valley to the Panhandle. Periodic reports from Washington and Chicago gave the program a national perspective. A staple to rural Texas for more than forty years, Hoover's Farm Bureau Roundup has become the longest running radio farm program in the nation.

Under Bill Hoover's leadership, "Farm Bureau Roundup" was originated, eventually going out to more than a hundred radio stations across the states.

Typical of the "Roundup" team members was Bob Walsh, who entered agricultural broadcasting in the mid-1960s after careers in the military and the Extension Service. Walsh eased into broadcasting through Extension-related television appearances in the Fort Worth and Dallas areas, where he caught the eye and ear of WBAP Radio Farm Director Bob Ethridge. About that same time, he also attracted the attention of Charley Manly, newly appointed director of the TFB radio "Roundup."

Walsh, in turn, was attracted to Farm Bureau's positions on political and production issues, though his espousal of such policies sometimes brought him unwanted attention from the station management. One frequent Walsh target was the Sierra Club, which earned his enmity by inviting him to speak at one of their meetings, then throwing him off the stage when they didn't like what he had to say.

Texas Farm Bureau, with modern equipment, provides video farm coverage to Texas broadcasters.

The incident, as he described it, became somewhat humorous in retrospect. It began when Walsh was invited by Dallas lawyer and ardent environmental activist Ned Fritz to address Sierra Club's North Texas chapter. Walsh said he was doing fairly well with a series of slides documenting insect damage to crops and the need for pesticides to deal with them. Then he came to a slide depicting a trapped coyote. The predator had killed fifty goats in a single night, he explained, but the Sierra Club crowd's reaction was "oh, that poor coyote." An SMU professor "shoved me off the rostrum," Walsh remembered, and "some woman got after me with her umbrella."

Telling such stories on the air earned him hate mail and threats, as well as complaints to his bosses. "I got chewed on quite a bit," he said. "I don't know how I kept my job because you're not supposed to editorialize, and I did it constantly."

Under J.T. Woodson's presidency, TFB added television to its Information Department in the fall of 1972, when it hired a young Waco newscaster named Patrick Batts to provide spot news and agricultural features for television stations statewide. Within two years, Batts moved on to the American Farm Bureau Federation, and by 1977 he assumed the post of information director for AFBF. The program he began at TFB continues to provide video farm coverage to Texas broadcasters, operating today with a technical sophistication that wasn't even imagined in its early days.

Closely related to the Information Department was TFB's Research and Education Department, created in 1958 and headed for decades by Bill Wedemeyer. Wedemeyer signed on with Farm Bureau in July of 1951 as one of its first field men. He soon developed a twist on the usual organizational work assigned to field men; in addition to coordinating membership drives, "selling" Farm Bureau in the hinterlands and helping county officials with record keeping, he began conducting county-level training programs intended to help local officials understand and deal with actual ag-related problems.

"The idea was to surface correct facts with regard to basic farm and ranch problems and distribute them to Farm Bureau leadership," Wedemeyer explained to an interviewer decades later. It was an idea that caught the eye of Roger Fleming, then secretary-treasurer of the American Farm Bureau Federation. Fleming summoned Wedemeyer to the 1957 AFBF convention in Chicago with the idea of luring him away from Texas and into the national organization's Washington office.

J. Harrell West had been elected earlier that year to replace the ailing Walter Hammond as the first new Texas Farm Bureau president in almost two decades, and he got wind of Fleming's plan before it could be executed.

"As we walked through a revolving door," Wedemeyer recalled, "Mr. West told me that he wanted to see me before I talked with Roger. Well, naturally, I wasn't interested in going to Washington if I could stay in Texas. West asked me to consider moving to Waco and establishing a research and education program for TFB."

That was the beginning of Wedemeyer's *Facts For You* newsletter.

Wedemeyer's first assignment was to study the idea of a state sales tax, an issue before the Texas Legislature at the time. It was such a hot topic within Farm Bureau ranks that TFB directors had voted during their 1957 convention to study whether such a tax would be good or bad, even before they had a department to conduct the study. The conclusion of that first study was that a sales tax by its very nature is progressive and "has no limits;" Wedemeyer distributed his report to county offices, but the tax was eventually implemented despite TFB's best efforts.

The *Facts For You* newsletter was a success from its initial issue produced first by Wedemeyer, then later by G. Glen Jones.

"My experience of seven years as a field representative," Wedemeyer explained, "taught me that if the program was to be successful, it must deliver facts to the ... leaders. It wouldn't do any good to stack facts up in Waco in a library and leave them there." He credited the idea of a newsletter per se to fellow early-fifties field rep R.J. Buchanan, who "encouraged President Hammond to designate someone to read the many reports and share a summary with field representatives." Buchanan's plan wasn't followed up at the time, Wedemeyer said, but it "stayed with me through the years."

The newsletter concept wasn't universally accepted within Farm Bureau, however, at least at first. Most people in Waco at that time felt like it was unnecessary, that there wasn't that much information that needed to go to the counties. But Wedemeyer persisted, and the first newsletter went out on May 9, 1958, under the tentative title of *Recent Developments.* By the end of the month, its name had changed to *Facts For You,* a moniker for which Wedemeyer gave much credit to TFB Information Director Bill Hoover.

The "weekly sheet," as Wedemeyer called it, was originally intended to go to county Farm Bureau presidents, state board members and field representatives. He never intended it to go to the public, in part because the "facts" his department unearthed and printed in the newsletter didn't always support Farm Bureau policy.

Counties were invited to subscribe on behalf of those leaders they thought would benefit, and "some of the counties wanted to send it to all their members." Wedemeyer disapproved, but despite that and the cost, at least a couple of county Farm Bureaus put their entire memberships on the subscription list.

The publication also went to selected libraries, including the University of Texas, Texas Tech, and the Texas Collection at Baylor. From the first, Wedemeyer's newsletter met with a mixed reaction in Waco. It was popular with the staff, he recalled, but not always with upper-level administration, "in that basic facts and not specific policy was emphasized."

Support from the board and "county people" was "strong enough that it maintained the publication," but there were some sticky moments. Wedemeyer remembered that Nueces County leaders became "terribly upset" over a study and report on the Federal Cotton Promotion Law. "That one was regrettable," he said, "but nevertheless, the facts were there. This was obviously a federal takeover of the cotton promotion program."

Wedemeyer credited education by his department with defeat of the 1963 wheat referendum, a "regimented, socialist type of program" that producers didn't like once they understood it. Not everyone wanted it understood, he suspected, recalling an incident in McLennan County when an ASCS committeeman denounced him from the floor of a meeting.

"We are not going to allow this young man to come out here and tell us how to vote," he remembered the man saying, to which he responded that "I only want you to understand the program and vote your honest opinion."

That was his standard position: he often held strong opinions on issues and expressed them freely, then encouraged members to make their own decisions. It was his job to see that they did so with all the facts at hand. To some West Texas detractors who agreed with what he had to say but criticized him for saying it, Wedemeyer answered, "I don't mind you going to Hell, I just want you to know you are on the road, so when you get there you won't be surprised."

The TFB Research and Education Department was not always embroiled in intramural controversy, however. Far more often, it provided

Farm Bureau leaders ammunition for battles that united the membership against common enemies. One such battle pitted Farm Bureau against the federal government over the Occupational Safety and Health Act, the OSHA enabling legislation.

Wedemeyer's newsletter warned in early 1971, within months after passage of the Act and even before the official date of implementation, that it "contains broad authority which will permit the closing down of businesses that do not comply." Typical of virtually all sweeping federal legislation passed within the last half-century or so, the OSHA bill left most particulars up to unelected and unaccountable bureaucrats to impose by regulation; no one really knew, months after passage, exactly what the bill meant. But TFB had a good idea: "It appears that regulations will cover any farm employing help that is not family related," Wedemeyer wrote, "whether full time or temporary."

Texas Farm Bureau never took workplace safety lightly. Accordingly, TFB maintained its own Safety Department for many years beginning in 1956, dedicated to reducing on-farm hazards, injuries and deaths. Its focus, however, was always on practical measures, such as the triangular "slow-moving vehicle" emblem, which the organization was instrumental in developing during the late 1960s. OSHA, by contrast, led to the employment of an army of bureaucrats who spent their time drafting regulations so hair-splitting that the specifications on ladders alone comprised a volume large enough to dwarf most small city telephone directories. Other tools and facilities were dealt with in equally extreme detail, in some instances down to the color of paint to be used on pillars and posts.

OSHA's sweeping regulatory power met with considerable opposition, and critics from throughout the productive sector immediately sought to rein it in. It was the job of Wedemeyer's department to keep abreast of what rules did and didn't apply to Farm Bureau's members, and to pass that information along to county leaders and the Information Department for dissemination. TFB also tracked the success — and failure — of reform efforts.

It was far from the first such assignment for TFB's Research and Education Department, and nowhere near the last.

Texas Farm Bureau: Knocking on Legislative Doors

The Texas Farm Bureau began to jell as an organization when it began offering such services as insurance and marketing assistance, but no concern is more important to agriculture than its treatment under state and federal government. It is in legislative and regulatory affairs that any farm organization truly earns its dues, and the importance of those programs grows by the year as government's power and reach expand.

Texas Farm Bureau had irons in the fire from its beginning, contributing input into the various Agricultural Adjustment Acts of the Depression years, sending Cliff Day and other representatives to Washington for the "Farmers' March" of 1935, and lending its weight to the "full parity" demands of the late Forties and early Fifties.

Along the way, TFB worked to promote rural electrification and was a major force in winning the Farm-to-Market road system for rural Texas. Subsidy programs, marketing orders and other pocketbook issues notwithstanding, the electrification and rural roads issues were arguably the most important gains made by agriculture in their era.

Rural electrification is one of the positive lasting legacies of Franklin Roosevelt's expansive New Deal era. Prior to the Rural Electrification Act of 1936, the electrical power that virtually all Americans take for granted today was unavailable to rural areas. Utility companies fought tooth and nail for franchises to serve population centers, but they could not be induced to extend the same service to sparsely populated regions. Their reasoning was simple and, from a practical standpoint, unassailable: they simply could not afford to string poles and wires far into the countryside, where profits and losses had to be calculated on the basis of miles per meter rather than meters per mile.

As a consequence, farm and ranch families struck matches to coal-oil or Coleman lanterns rather than flipping a light switch on the wall, and they heated and cooked with wood, coal-oil or butane. Where they could pull in a clear station, they listened to Lum and Abner on battery operated radios, keeping their fingers crossed that the batteries would last at least as long as the program. Some fortunate country folk could afford wind chargers or motorized generators, and some fewer could keep them operating, but even those costly marvels compared poorly against the plentiful electric power their city kin accepted as their due. Electric milking equipment, feedmills and even shop tools were out of the question.

Food preservation was more carefully practiced art than easily dismissed science; refrigeration depended on blocks of ice that were apt to arrive in anything but prime condition after long trips from town on poor roads. For lack of a better or more reliable method, many farm and ranch wives still cooled milk in the flow of water from a spring or windmill, and canning was the preferred method of storing everything from fruits and vegetables to prepared meats. Fresh meat hung from a rafter in the shade, covered with a cloth to keep the fly-blows to a minimum, and it had to be consumed in a matter of days lest it spoil — in the summer it was a matter of very few days.

Those conditions, of course, were so common — and in many ways such an improvement over what had been the norm only a few decades earlier — that there was nothing especially remarkable about them. It had always been that way, or worse, so there was no universal condemnation of the situation. There was, nevertheless, a steady push for rural electrification, which intensified as the disparities between town and country dwellers increased.

What really brought the matter to a head, ironically, was not the perceived need of rural Americans for electricity so much as the perceived need of other Americans for the jobs that would supply rural electrification. Such politicians as Gifford Pinchot, George W. Norris, Morris Llewellyn Cooke and Franklin Roosevelt had championed the cause of rural electrification since the Teens, mostly in vain. Even with Roosevelt in the White House, their campaign was an uphill battle. Despite intense pressure from the American Farm Bureau Federation and its state affiliates, including Texas, rural electrification was little more than a dream.

Until, that is, the Emergency Relief Appropriations Act of 1935. In the interest of putting Depression-plagued men to work, that act authorized the expenditure of up to $100 million for rural electrification. The rationale for the project was not only that it would create jobs for electric linemen, but in the timber industry for poles, the wire industry for conductors, and among suppliers of insulators, transformers and other equipment. Once underway, it was also expected to create demand for electric appliances and even such mundane items as lightbulbs.

It wasn't until the following year, with passage of the Rural Electrification Act, that wiring the countryside became a goal in itself.

Almost immediately, however, the new Rural Electrification Administration became snagged on procedural problems of the sort that so often arise when governments set out to solve a problem without thinking it through first.

In the process of creating and unleashing the New Deal's many alphabet relief agencies, FDR attempted to target much of the massive funding toward employing the unemployed, to manpower rather than machinery, to those on relief rather than those who already had jobs. To that end he decreed that at least a fourth of all relief program funds be spent directly on labor, and that such labor be recruited from the ranks of the jobless. That made sense when the project in question was a crude rock wall in some public park, but it ran into trouble when the project was a powerline, generating plant or substation; that type of construction required skilled labor of the sort that even in the Depression was not likely to be hanging around a relief office.

Rural electrification became one of the major problems for country dwellers trying to dig themselves out of the Depression.

It took months to meld intent and outcome, even with Roosevelt's eager support, and in the end, the most practical solution called for shifting REA's focus from grants and direct subsidies to low-interest loans. The latter were not subject to the well-intentioned but hopelessly complicating restrictions.

The problem then became one of generating demand for the loans. Early rural electrification plans had envisioned enticing commercial power companies with government assistance, but those hopes quickly faded, at least in part because of REA's shift from outright grants and subsidies to loans; money that had to be paid back was not nearly as enticing. As a result, REA officials began reaching out to cooperatives and nonprofit corporations.

This would have two unfortunate and self-reinforcing consequences. Commercial utilities rankled at the notion of competition, even for rural business they didn't want. Moreover, efforts to blaze new ground for cooperatives and non-profits at a pace that satisfied REA's "emergency" mandates created a mushrooming bureaucracy that legitimately frightened

The fight for rural electrification ultimately brought backroads Texas into a modern era of electricity and communication.

many Americans.

Within a year's time, the Rural Electrification Administration grew from wet ink on a Presidential signature to an independent federal agency with a plethora of departments. In addition to divisions devoted to such practical pursuits as improved engineering, REA boasted a "Utilization" division, which in essence promoted electrification among rural people who might not yet know they wanted it; an "Examining" division that helped fledgling cooperatives write applications it would then approve; and a "Division of Operations Supervision" whose goal in large measure was to help cooperatives build membership. By 1939, almost eight hundred employees were on REA's payroll, not counting those of the scores of cooperatives it had helped create by then.

Each and every division was legal, and their functions were consistent with the law's expressed intent of expanding rural electrification as quickly as practical. In an America not yet inured to bureaucracy, however, there was something unsettling about an agency that seemed so boldly constituted for its own perpetuation and expansion. REA's commercial utility rivals lost no time in capitalizing on that image, casting the entire rural electrification program as a conspiracy to create a vast federal power system. It was a convincing argument, and even some of rural electrification's strongest supporters began to have doubts about particular aspects of REA's activities.

Farm Bureau was among them; though still staunchly behind total electrification of rural America, AFBF and some of its state affiliates — including the Texas Farm Bureau — began openly questioning just how that goal was being accomplished and what the taxpayers' money was being spent on. Those questions provoked a swift and bitter response from REA and the co-op interests, which accused Farm Bureau of conspiring with commercial utilities to destroy the very program they had been instrumental in creating. The enmity would last for years, breaking out every decade or so in open skirmishes.

Nevertheless, Texas Farm Bureau's dedication to rural electrification remained undimmed, as evidenced by its prominence throughout the years on the organization's legislative to-do list. Walter Hammond and his successors seldom failed to mention this or that battle on behalf of "REA" in their annual convention addresses, and Hammond often chronicled the program's annual progress toward the goal of total electrification.

Typical were Hammond's remarks during the organization's 1944 convention. Noting that Farm Bureau "sponsored legislation that made possible REA," he vowed that the group would "be on hand with plenty of power to defeat proposed rules and regulations that would — if enacted — seriously stifle the expansion of REA."

That address also brought to the fore the first tentative steps toward what became Texas' nationally renowned farm-to-market road system. Hammond mentioned "a new plan for farm-to-market roads" that would earmark certain funds for that purpose once World War II was over. There was a catch, however — a big one. The funds to be earmarked would be raised by repealing the tax exemption on gasoline used for off-road agricultural vehicles, a provision to which Hammond was unalterably opposed.

Those who had apprised him of the proposal "explained that it was the thinking of the group that the farmers' first reaction would be to oppose the plan," he told Farm Bureau conventioneers, "but that after the plan was explained and understood, it would be accepted by the farmers.

"Well," he continued, "I am a farmer, the plan was explained to me and I think I understand it, and I will give you my reaction — When the tax was placed on gasoline, it was agreed that people who travel the roads should pay for the construction and maintenance, and I agree with that — so the tax was placed on gasoline burned in motors used to propel vehicles on the roads and highways of Texas. Later, it was discovered that around $12 million of this tax was refunded to non-highway users of gasoline, with about $8 million of the twelve going to farmers, which seems to annoy some people.

"Now, for several years, continued efforts have been made to get that $8 million from the farmers. Several schemes have been tried. Now this one — and I am not deceived by this move to 'help' the farmers by having them pay four cents per gallon on gas used in tractors.

"Farmers are paying four cents per gallon on (vehicle) gasoline while most of their travel is on dirt and mud roads, and if anybody wants to build roads for the farmer, then do him a real service by using at least part of the farmer's gas tax to build farm-to-market roads."

Hammond added a clear warning to those promoting the tax exemption repeal scheme: "If such a bill is introduced in the legislature, I will round up the farmers and start a row that will make the folks in the recent Texas political row look like a covey of tired old men."

A year later, things were looking up. World War II was over, and Texas Farm Bureau had indeed beaten the proposed repeal of the off-road fuel tax exemption. With it, however, had gone the farm-to-market road program — at least for a while.

In the late 1940s, Texas Farm Bureau was trying to secure legislative funding to "pave the last mile" and "get the farmer out of the mud."

By the time the biennial Texas legislature reconvened for its 1947 session, the Texas Farm Bureau had drawn something of a line in the sand on the rural road issue: not only would it oppose any alteration in the gasoline tax situation, it also favored vesting authority for farm-to-market roads in county commissioners' courts rather than in the Texas Highway Department. The latter was a matter of economics, as Walter Hammond and Farm Bureau saw it. They reasoned that more miles of caliche and gravel roads could be constructed on a given budget than asphalt pavement, and opted for the miles. That fit the programs of most county road departments, whereas the state highway department favored true "all-weather" pavement. It was a plan to "pave the last mile" and "get the farmer out of the mud."

A *Texas Agriculture* editorial at the close of the 1947 session lamented the failure of "many attempts ... to pass legislation for the purpose of building farm-to-market roads," and placed blame squarely at the feet of "the cement and asphalt lobbies." Hammond accused those interests of opposing any plan that did not center on the use of their products.

Farm Bureau regrouped and used the next year and a half to draw up its own bill in cooperation with the County Judges and Commissioners Association and the Rural Letter Carriers Association. In its initial versions it was a bold and ambitious plan, calling for a "minimum" of sixty thousand miles of new road construction within the first five years, and seeking $40 million from "surplus" state funds in the first of those five years alone. Later years were to be funded with a "natural resources" tax that TFB estimated would generate $27,500,000 annually. Hammond gave some ground in his opposition to administration by the Texas Highway Commission, but the proposal sought to save the idea of constructing more miles of lesser roads by reserving to county commissioners the authority to determine what types of construction would be used on roads within their jurisdictions.

Texas Farm Bureau, led by President Walter Hammond (standing center) and Vice President Frank Taylor (standing left), took their fight for farm-to-market roads to Governor Allan Shivers at the state capital.

Introduced as House Bill 59 by State Rep. C.S. McLellan of Colorado County, the actual legislation differed from the interim plan in several particulars, including abandonment of the goal of capturing tens of millions of dollars in "surplus" funds to jumpstart construction. The vaguely defined "natural resources" tax was also focused and pared down to a "privilege" tax of one cent per thousand cubic feet of natural gas produced within the state. Mindful of the fact that many Farm Bureau members benefitted from oil and gas production on their land, the bill's backers deliberately exempted royalty owners from taxation. To sweeten the pot, the measure called for a fourth of the tax revenue to go to the school fund.

HB 59 met almost instant resistance, both direct and indirect, from what Hammond dubbed "the multi-million dollar pipeline interests." A succession of House votes delayed consideration of the measure, and another trimmed twenty-five percent off the proposed penny levy. Within a month after its filing, Hammond and *Texas Agriculture* were proclaiming the McLellan bill "endangered" and exhorting Farm Bureau members to wire and write their state representatives in support.

"If you want a road bill," Hammond warned, "you must act — and act fast! Farm Bureau is in the midst of the toughest fight in which it has ever been engaged."

As the *Texas Agriculture* reported, "Some 3,000 poll tax payers from the cross roads swarmed the capitol on March 7 to see the Revenue and Taxation Committee report their rural road bill favorably....For years the lawmakers have been telling us they would pass a measure to build that last mile of road if we'd agree on a plan. The sons of sweat and toil—the farmers and ranchers, the rural letter carriers, the school teachers, the commissioners and county judges, the small town country editors and business men—have been working two years (since the last legislature) on a road bill on which all could agree.

"After telling the boys in Austin they were agreed on the bill, the farmers went back home to plow and the letter carriers to deliver the mail, confident that their representatives would remain loyal to the folks that put them in office. But the multi-million dollar pipeline interests, who can spend millions to defeat a single natural resource tax measure, stayed in the halls of the capitol at Austin and they have let no grass grow under their feet. Proof of this is one delaying action after another on the McLellan Bill, and the vote on Monday to whittle the tax rate from one cent to three-fourths of a cent per 1000 cubic feet of natural gas, is their handiwork."

Meanwhile, opponents introduced rival legislation known as the Colson-Briscoe bill. Those favoring the Colson-Briscoe Bill, however, said that a vote against the McLellan Bill was in no way a vote against the farmers. Most farmers did not believe them. The *Texas Agriculture* explained:

"The Colson bill is dependent upon appropriations out of the general fund and will be used on government specified hard surface roads. The bill provides only $5,000,000 more a year than we have been getting for that purpose, which will give farmers only two more miles of road per county each year.

"The McLellan bill is designed to build feeder type rural roads out of gravel, caliche, shell, etc., where practical and hard surface roads in counties where cheaper materials are not available or are impractical. The McLellan Bill is designed to build "that last mile." TFB also warned that the Colson-Briscoe measure would not provide a "permanent" contribution to the rural road problem, as its backers insisted.

However, Hammond, in an editorial, did say, "In fairness to Colson and Briscoe, I am sure they intended that the legislation would be permanent, but not so with those who would do favors for the natural gas industry."

Those warnings went unheeded; the Colson-Briscoe bill passed and the McLellan bill ended up in the trash bin.

When the 1951 legislative session opened, Texas Farm Bureau and its allies were back with a new variation on the by-then familiar theme. That one, House Bill 669, sponsored by Jim Sewell of Blooming Grove in Navarro County, revived the penny gas levy, this time as a "gathering" tax on gas entering pipelines. Some inter-session study had revealed that the

majority of Texas-produced gas left the state untouched by any tax —one figure cited in *Texas Agriculture* said sixty-two percent, another seventy-five percent — and backers evidently felt that building Texas roads at the expense of "northern industrialists" and other out-of-state interests would appeal to voters as well as their legislators.

The Sewell bill set aside money for bridge construction. Because of bad rural roads, milk trucks and mail carriers would not cross dangerous bridges. But school buses had no choice.

The quarter-cent share of funding designated for education remained intact, and to broaden its constituency, the bill set aside another twenty-five percent of revenues for street and bridge construction in cities and towns, leaving just half of the levy to fund rural roads. Ironically, given that the bill was estimated to generate about $30 million per year, the dollar total anticipated for farm-to-market road construction would only be $15 million, the same as that provided in the Colson-Briscoe bill.

One difference was that the Sewell bill would direct its revenues to the counties, where TFB and other backers remained convinced that the dollars could be stretched with less expensive construction methods. The other difference was that they still considered the Colson-Briscoe measure could be subject to repeal at any time. In a classic example of political obtuseness, Farm Bureau's opponents gave them a tremendous boost by proving them right on that count; in mid-February, with the 1951 session only weeks old and barely past the mandatory warmup period, State Rep. Callan Graham introduced HB 226, which not only would have repealed Colson-Briscoe but would have raided county treasuries for additional funds.

How much sympathy mileage the Sewell forces made off the Graham attack is impossible to say, but the bill was so poorly received that House leaders, spearheaded by Dolph Briscoe, actually lifted it out of a "friendly" committee and reassigned it to the House Agriculture Committee, where it was guaranteed never to see the light of day. And weeks later, the House passed the Sewell bill, the first time in decades of trying that the Texas House voted to levy a tax on natural gas. Even more remarkable, the bill passed without a single significant amendment.

The climate in the Senate was considerably cooler. Sponsored there by Sen. George Nokes of Corsicana, who had held Sewell's House seat before election to the upper chamber, the bill was reported out of the Senate Agriculture Committee in early May and then rebuffed two to one on the Senate floor as an amendment to an omnibus tax bill. That bill, sans Sewell-Nokes, went through conference and was then soundly rejected by the House. A second conference committee compromised with parts of the Sewell-Nokes legislation, and both houses accepted that final bill.

The *Texas Agriculture* reported: "The people of Texas won a great victory in the session of the Texas Legislature, which has just adjourned.

"In the first place, the actions of the Legislature made it certain that there will be more money available for the building of farm-to-market roads than ever before.

"In the second place, a substantial tax was placed on the natural gas which is going out of Texas to thirty other states and two foreign countries. For more than twenty years, a fight has been made to tax this natural gas, but this year was the first time that these efforts have been successful.

"The Farm Bureau has fought long and hard for these two objectives—to build more farm-to-market roads and to tax the natural gas leaving the state—and the organization played a major role in bringing to pass these great successes this year."

While the measure did not pass as Texas Farm Bureau had written it, the battle undoubtedly prevented not only a severe cut in rural road funds but also the enactment of additional sales taxes. The organization appreciated the diligent efforts of Senator George Nokes and Representative Jim Sewell, who had been blinded while serving on the carrier USS Hornet during World War II. And Farm Bureau applauded the leadership of men like Representatives Dolph Briscoe of Uvalde, Clyde Whiteside of Seymour, D. B. Hardeman of Denison, Douglas Crouch of Denton and Otis Henderson of Waelder, among those who sidetracked the Graham measure and fought for the passage of the Sewell bill.

D. B. Hardeman wrote *Texas Agriculture*: "During the last session of the Legislature...a bunch of legislators were sitting around one night talking about our fight to save the rural road program. Included in the group was Rep. Maury Maverick, Jr., of San Antonio—the only big-city legislator who always voted for rural roads—and his father, former Congressman Maury Maverick.

"We were complaining because the big-city legislators were practically solid against us.

"'They just don't know what they're doing,' former Congressman Maverick broke in. 'They don't remember that Rome fell because it failed to build farm-to-market roads.

"'Rome placed all its emphasis on building great cities, and on great express highways linking those cities. It completely neglected the small towns and the farming areas...The farmers couldn't get their crops to market, and at times the people in Rome and other cities didn't have enough to eat...

"'At the same time the people in rural areas couldn't get to the big cities to trade. Lacking good roads, more and more people left the impoverished farms to live in the cities. This resulted in mass unemployment, turmoil and misery in the cities.

"'This weakened the whole structure of the Roman Empire. Agriculture went to pot, the cities were seething with hungry and unemployed people, the whole Empire rotted, and one day the barbarians pushed the whole structure down. That whole thing could have been avoided if the Roman emperors had built those rural roads which would have kept agriculture prosperous—roads linking the countryside with the cities.

Texas had a lot of land. But it took the Colson-Briscoe bill, then the Sewell bill, to bring roads to the state's rural corners.

Through the years, Texas Farm Bureau's relentless battle to bring good roads to country farmsteads finally paid off.

"'That one example in history ought to be enough to make every big-city legislator in Texas work for farm-to-market roads just as hard as you boys from the country.'"

Eventually enough of them did.

Texas Farm Bureau didn't get all it wanted — most of the money earmarked for rural roads would still be spent on more expensive hardtop construction — but it saved the tenuous Colson-Briscoe funding, with considerable additions. Farm Bureau also won what it had come to consider a moral battle with the "multi-million dollar pipeline interests." Meanwhile, farm-to-market road advocates had also defeated diversionary bills that threatened to raise the gasoline tax, which Farm Bureau for years had been criticizing as unfair enough already because so much rural driving was done in dirt and mud using gasoline taxed to pay for others' highways.

All in all, Farm Bureau considered it a good fight and a significant victory.

Farm Bureau's next big battleground was on the labor front. Agricultural labor relations had been a touchy area since at least the Depression years, as exemplified and personalized for the general public by Steinbeck's *Grapes of Wrath.* Such dramatic renderings of the migrant laborer's lot aroused considerable public sympathy, which television was able to exploit later with exposé-style "documentaries." By the 1960s, militant farmworker unions were openly confronting agricultural producers and skillfully managing the media coverage to their own benefit.

California was the unquestioned hotbed of farmworker activism, but Texas was not far behind, owing in part to its position as the nation's largest supplier of migrant labor.

High-profile issues used to capture the public's attention ranged from unsafe transportation, dishonest labor agents and poor housing to child labor, educational gaps and spotty health care, conditions that tended to be inherent in a nomadic lifestyle. Other citizens as a general rule were expected to address those problems themselves, but advocates among the union movement, liberal elites and some church groups demanded different rules for farmworkers, who they insisted were helpless when it came to defending their own rights. Such an attitude smacked of paternalism at best and bigotry at worst, but in some circles it was considered a serious breach of etiquette to point that out.

Down in the trenches, meanwhile, the real battle was over such prosaic issues as the minimum wage, workers' compensation and unemployment insurance. Farm Bureau, led by president J.H. West, opposed extension of those programs, primarily on the grounds that agricultural producers, unlike industrial or service-sector employers, cannot readily pass such increased costs on to their customers; the farmer or rancher is a price taker, not a price setter.

Advocates of dramatic changes in the farm labor arena gave lip service to those concerns, as exemplified by a 1960 pamphlet printed and distributed by the "National Council on Agricultural Life and Labor," a Washington lobbying group. The pamphlet calls for a slew of federal laws regulating agricultural labor, and takes pains to differentiate between "family farms" and "agribusiness." The latter, it says, amounts to "just five percent of all farms," are "corporatively owned...by absentee landlords," and can be found "sprawling over thousands of acres." The clear implication is that the laws being championed would apply only to those big "agribusiness" operations, when in fact most of the ag labor legislation promoted through the years has made little or no distinction between "family farms" and "agribusiness." Farm Bureau policy made no such distinctions because the members writing policy knew they would all eventually stew in the same pot.

As the years went by, Texas became more and more an urban state, its legislators and judges increasingly distant from agriculture. That favored the unions and their supporters, who began gaining ground legislatively and, of far more sweeping significance, judicially. Some of the ploys used to manipulate public opinion would have been laughable had they not been so serious.

One legislator in the early 1980s actually called a press conference on the Capitol steps to unveil a bill intended to ban "short-handled" hoes. Its intent was to portray agricultural employers as heartless slavedrivers who relished watching their peons stoop and bow. Farm Bureau pointed out that the hoes in question were used only in specialty applications such as ornamental horticulture, and the bill went nowhere. It had, however, doubtless left one more in a long string of indelible mental images with many malleable souls.

And those images eventually took their toll. By the mid to late 1980s, activists were scoring limited victories in the Texas legislature, and what they couldn't get from lawmakers they managed to get in court. The last of the big dominoes began to fall, and Farm Bureau was forced to swallow some bitter pills, among them workers' compensation and unemployment insurance. All the organization could really do by then was attempt to influence the details of legislation, to temper the new laws and make them at least survivable.

In the process, Texas Farm Bureau, rallied by presidents Carrol Chaloupka and S.M. True, became deeply involved during the 1980s in the struggle to reform a Texas workers' compensation program that had grown increasingly burdened by self-serving attorneys and outrageous jury awards. It was a battle that would have been almost unthinkable a decade earlier.

Another of Farm Bureau's major battles with government involves pesticides. Some early organic insecticides such as nicotine sulphate have been around for a century or more, but the "modern" age of pesticides really began with DDT prior to World War II. Applied to property, crops and people alike, DDT was credited with saving the lives of literally millions of people during the war years by fending off disease-carrying vermin and by protecting critical food supplies, both growing and in shipment.

That general-purpose pesticide and the improvements that followed helped bring about the "green revolution" that had allowed fewer people working fewer acres to produce more food than the planet had ever provided before. Along with improved crop varieties, mechanization and storage techniques, pesticides have embarrassed a whole series of flat-earth soothsayers predicting global famine. The oracles of doom, however, have a deep capacity for such embarrassment; when one approach doesn't work, they try another and another.

Author Rachel Carson chose to attack the source of the embarrassment itself, throwing down the gauntlet with her 1962 book, *Silent Spring*. The book's main contention was that pesticide residues were building up in the environment after only a few decades of use, thus posing a threat to the survival of all life.

Though much of it has since been demonstrated to be groundless — certain "man-made" chemicals found in Antarctica and assumed to be pesticide residue turned out to be a natural byproduct of red seaweed, one of the most plentiful plants on the face of the earth. Still, Carson's apocalyptic vision of a planet devastated by toxic compounds frightened enough of the public to generate pressure for massive new pesticide regulations. The upshot was the Federal Insecticide, Fungicide and Rodenticide Act, which created the U.S. Environmental Protection Agency. Staffed from the outset with ambitious young activists, EPA made DDT one of its first casualties, then set its sights on a wide array of other chemical compounds.

A pattern soon emerged; activists on the outside would sue their comrades on the inside, seeking cancellation of a pesticide or some other goal they could not accomplish through legislative action. Indulged by a friendly judge, the activists on the inside would "roll over" and agree to conditions that in many instances they lacked the authority to approve. In this way, EPA's powers soon expanded well beyond the scope of the original FIFRA. The arrangement also benefitted the organizations which brought the suits, in that EPA's acceptance of the lawsuits' terms often left the government holding the bag for the plaintiffs' legal expenses. The Natural Resources Defense Council was but one of several such groups of lawyers who supported themselves at the expense of hapless taxpayers.

Farm Bureau, both in Texas and nationally, sought to rein in some of EPA's sweeping authority by imposing limits on the agency's ability to issue arbitrary regulations. TFB's role in that effort was critical because of the group's cordial working relationship with U.S. Rep. Kika de la Garza, the Mission Democrat who has long wielded power in the House Agriculture Committee. Through the years, Texas Farm Bureau had been more effective, had exercised a more influential voice, in Congress because of that strong relationship that President S. M. True had built with Representative de la Garza.

In fact, Rep. Kika de la Garza personally spoke at True's retirement reception, expressing his admiration for the man and the organization he had represented so effectively. De la Garza said, "I was in Corpus Christi earlier today and had a difficult time getting here, but I would have come around the world ten times to be here tonight.

"All I can say is that in behalf of agriculture as I represent it on a national level, on the world level, we appreciate the impact, the very positive impact that this gentle and kind man has had on agriculture.

"I want also to say that there is an old Spanish saying about an individual who knows people, who treats people. Translated, it is the gift of the people, one who is able to not go above and not go below, but work with people.

"When the original Spanish explorers came across they were resting one evening by the side of a huge mountain, and one of them asked, 'Will anyone ever know that we came this way?' So the leader took a saber and carved his name, adding the words '*paso aqui*,' or 'came this way.'

"So I would say that Farm Bureau, Hale County, Texas, the world is better because S. M. True '*paso aqui*.'"

Over time, EPA's rulemaking process became more deliberate and less freewheeling; regulatory proposals were shepherded step by step through a lengthy string of drafts, hearings and comment periods. That minimized the opportunities for gung-ho bureaucrats to ramrod a set of rules through without opponents' knowledge, as had been common in the early 1970s. By the same token, however, critics of some of EPA's early roughshod work found themselves unable to reverse bad policy. The rulemaking procedure was equally strict in both directions, and opponents of a particular regulation often found themselves faced with the impossible task of disproving a nebulous "fact" that had never been demonstrated in the first place.

One Farm Bureau goal that remains elusive despite almost three decades of work is forcing government pesticide regulators to balance perceived risks of a given compound with its known benefits. In instances where EPA has been held to something approaching that standard, the agency has often minimized the risk/benefit conflict by pointing to the availability of alternative products while in fact proposing the "alternative" compound for cancellation as well.

The pesticide struggle is far from finished. On the one hand, EPA's aggressive attacks on existing chemicals have forced scientists and manufacturers to search out new, often more specific alternatives. On the other hand, those alternatives can be much more costly than the products they replace, at least in part because of EPA's own stringent registration process; more than a few promising products with genuine environmental benefits have been abandoned in the trial stages when their makers realized they could never recover registration costs.

Closely akin to the fight over protecting crops from insect and rodent pests is the battle to protect livestock from predators. This has mostly involved coyote depredation on sheep and goats, but economic losses also have been severe from such diverse species as fox, bobcats, cougars, bears, eagles, feral hogs and even ravens. Just as they did with pesticides, environmental activists have turned their attention toward predator control by banning most lethal control measures.

Crops are sprayed with a pesticide that allowed fewer people working fewer acres to produce more food than the planet had ever provided before.

Their excuses run the gamut, but most hark back to an insistence that predator damage actually benefits stockmen because it performs a culling function. Never mind that culling is the stockman's job or that when he culls an animal it usually returns at least some salvage value.

As with the pesticide battle, the war on predator control caught fire during the late 1960s. Hastily assembled "study" panels recommended elimination of all government support for predator control programs, and a somewhat laughable — in retrospect — effort by the Nixon administration to curry favor with the left during an election year led to a 1972 executive edict banning the use of predator toxicants on all federal grazing land. The as-yet unchained EPA joined in with a blanket predacide ban a few weeks later. In keeping with the activists' modus operandi, the EPA ban was a "settlement" of a lawsuit, and its terms were sealed so opponents could not learn what had happened until long after the deadline for appeals.

Farm Bureau and various livestock organizations were years discovering how they had been double crossed. Farm Bureau, meanwhile, backed an "interim" indemnification program to pay producers for increased losses as a result of the predacide ban. The organization intended its indemnification program to last only until federal researchers had identified and implemented new control methods designed to take the place of the banned toxicants.

Neither the indemnifications nor the promised studies ever materialized. The government funded some desultory research into such efforts as making sheep taste like toads and raising their aggression level with LSD. Just how psychotic sheep with a flavor too rank to interest a hungry coyote were supposed to remain marketable was never clearly explained. In the main, however, the activists who had justified their ban by promising research threw off their camouflage and lobbied directly for cessation of all lethal controls. Sheep, not coyotes, should be controlled, they insisted.

Farm Bureau and other producer groups fought a holding action throughout the remainder of the 1970s. Then came the Reagan administration. Just weeks after his inauguration, President Reagan lifted the Nixon executive order banning toxicants from federal lands. At that point the gesture was little more than symbolic, however, because of EPA's overriding predacide ban. It took a concerted effort by Farm Bureau and its allies over the next few years to see the ban overturned in court.

Back in the arsenal was the M-44 sodium cyanide ejector, a spring-loaded device which propels cyanide into the mouth of a coyote when the beast bites and pulls a scented trigger mechanism. The ruling, issued by an EPA administrative law judge and upheld by the federal courts, also ordered EPA to re-register single lethal-dose baits using Compound 1080 predacide. Added to the predator control arsenal was the livestock protection collar, a rubber bladder arrangement containing Compound 1080 which ruptures and poisons the coyote as he presses the typical throat attack.

Even with the ruling, it took years for Farm Bureau and its allies to maneuver the collar and M-44 through the bureaucratic maze, and more than a decade later EPA is still ignoring the order to register single-dose baits. Meanwhile, TFB has fought skirmishes to make protection collar rules more practical and to head off proposals that would impose mandatory husbandry practices as a prerequisite for predator control assistance.

A problem that had long faced both big producers and home owners alike was the attack of imported fire ants in rural and urban Texas. The fire ant had migrated from South America around 1918, arriving first in Mobile, Alabama, then spreading out throughout the South. By the 1970s, fire ants had infested 150 million acres in nine states. As Texas Farm Bureau reported, "Fire ants can be found in sixty-one Texas counties on approximately 23.5 million acres. The imported fire ant is a pest to humans, animals and crops. Its venom causes boil-like sores and has put many people in the hospital. It eats plant roots, stems, seeds, tender shoots and some insects. The ant's large, hard-crusted mounds damage farm machinery and ruin pastures."

The facts were grim. There was no such thing as eradication, people were being told. The best anyone could hope for was a good control program.

Pesticides containing Mirex had been the only effective weapon against the fire ants. But on March 18, 1971, the Environmental Protection Agency issued a notice of cancellation of Allied Chemical's registration of pesticides containing Mirex for the use against the imported fire ant. There were serious questions, the EPA said, about the effect of Mirex on human health and on animals.

Allied Chemical promptly challenged the order, petitioning to have the matter referred to a scientific advisory committee which could evaluate all relevant scientific evidence concerning both the benefits and risks from the use of Mirex.

The committee studied the evidence, then recommended that Mirex registrations should be continued with labeling restrictions to minimize environmental contamination. It also recommended publicly supported control programs that were limited to infested areas where the imported fire ant interfered with agricultural operations. For a time, EPA allowed limited federal/state program use of Mirex to provide relief in areas with large fire ant populations. Initially, rural Texas fought fire ants by spraying the pesticide. Now it was forced to use Mirex from mound to mound, even though ground application was found to be largely ineffective.

By 1973, William D. Ruckelshaus, EPA administrator, was holding a hearing to determine whether or not the registration of Mirex should be canceled or amended. Texas Farm Bureau immediately became a party to the hearing, testifying in behalf of those Texas counties, those Texas farmers and ranchers who daily had to battle the infestation of the imported fire ant.

In spite of a strong effort by Farm Bureau, EPA outlawed the use of Mirex.

Farm Bureau leaders traveled to Washington, D.C., on many occasions to testify in behalf of programs vital to the agricultural community.

Since then, Texas Farm Bureau has been instrumental in encouraging the state legislature to appropriate approximately $1.9 million annually for the control and possible eradication of fire ants.

Perhaps no issue, however, has galvanized the Texas Farm Bureau in recent years to the same extent as private property rights. Farmers and ranchers, after all, make their livings from the land, and many of them have been on the same piece of ground for multiple generations. It should surprise no one that the average Farm Bureau family would fight to defend its property just as the family members would fight to defend one another.

For years, that fight was one-sided and landowners were on the losing side. Time and again state and federal agencies trained their sights on property rights, picking off stragglers and building an impressive, if distressing, record of victories over landowners.

Texas Farm Bureau, for example, was successful in passing Texas' "Right to Farm" law in 1981, just before redistricting took a chunk of the legislative clout out of rural areas. The law forbids cities and towns from interfering with existing agricultural operations even as they spread out and engulf them. A loophole, however, requires that the operation be in compliance with applicable state and federal regulations, and the State of Texas has interpreted that to mean virtually anything that suits the needs or whims of regulators; if suburbanites move in on a feedlot, dairy or auction barn, for instance, and later decide they don't like the smell, the "Right to Farm" law would be helpless to protect the agricultural operation no matter how long the law had been in place.

In the last few years, however, the tide has turned against the regulators. Emboldened by their successes, state and especially federal agencies have been caught overreaching, and even town dwellers who never gave much thought to property rights disputes have begun to recognize and rebel against the threat. The last straw, it appears, is the federal Endangered Species Act, sold originally as a way to protect such high-profile "endangered" species as bald eagles and grizzly bears. Instead, the law has been used far more often to shut down dams and developments opposed by environmental activists.

Its promoters overstepped, however, when they began fining and imprisoning homeowners for protecting their property. Suddenly the ESA ceased to be an obscure law impacting "greedy" developers and rural landowners; it became a personal threat with which every homeowner could identify.

Texas Farm Bureau was in the vanguard of individuals and private organizations seeking reform of the Endangered Species Act years before reform became a popular notion. That leading role earned it credibility among non-members when golden-cheeked warblers, black-capped vireos and gnatcatchers made thousands of Central Texas retirement homesites virtually off-limits to their rightful owners.

Texas Farm Bureau had long been a staunch defender of the fifth amendment of the Constitution which provides that no private property may be taken from individuals without compensation. It was, TFB firmly believed, one of the cornerstones of the free enterprise system. When the government found it necessary to actually take title to private property, compensation should be a matter of course. Property owners were to be compensated, even if the actual value of the property might be disputed.

But now a new kind of taking had emerged because of environmental regulation. It was described as "regulatory taking," occurring when regulations—such as the Endangered Species Act, Wetlands or other regulations—were imposed. Federal and state mandates often prohibited management practices, reducing the profit potential of the property and even severely diminishing its value.

As far as Texas Farm Bureau was concerned, this situation definitely could be described as the taking of property rights, and that policy, by any means and for any reason, was undesirable.

Texas Farm Bureau would establish a Property Rights Coalition, working with its friends in the Texas Legislature on the specific language for property rights legislation. Bob Stallman, who became TFB's president in 1993, said the legislation would focus on three areas: protecting the value of property for property owners, seeing that common sense was put into the regulatory process where cost is determined versus benefits to society, and assessing government actions to see what impact they will have on private property.

Additionally, Texas Farm Bureau is supporting legislative amendments to block unfunded federal mandates, promote risk assessment and cost benefit of federal regulations.

Those three points were widely criticized by environmental groups that called them the "unholy trinity," particularly when they realized that the three formidable issues were putting the brakes on their agenda in Washington, D.C. In reality, they were simply an attempt by citizens to regain control of the federal government.

Farm Bureau leaders from Texas also traveled to Washington to testify before a U.S. House subcommittee, urging lawmakers not to approve a bill designed to increase the size of the Big Thicket National Preserve, pointing out that such expansion would cause economic harm to the timber industry and ecological problems to that area of East Texas. The proposal did not have the support of the National Park Service. And the Farm Bureau argued that passage of the bill could have severe implications for private property rights. From an ecological standpoint, the wilderness area would disallow sound management of insect and disease infestation, fire control and water quality.

The Farm Bureau stand, under the leadership of president True, was supported by Hardin County Judge Pete McKinney, who told the House subcommittee that East Texas had witnessed a decline in oil and gas, and that the other two major natural resources—water and wood—must be preserved. "We cannot afford to lose the water and wood," he said.

Farm Bureau also opposed a permanent injunction issued by U.S. District Judge Robert Parker of Tyler in June of 1988, preventing the U.S. Forest Service from using evenage management—a method of harvesting in which all trees within a tract of land are cut down and other plants bulldozed and burned—within three-quarters of a mile of red-cockaded woodpecker colonies in national forests. The injunction also required the closing of about one-third of the unpaved logging roads in the forests.

The ruling caused the suspension of eleven logging operations including seventeen million board feet of lumber. As many as 200,000 acres of national forests could ultimately be affected. The lawsuit caused the immediate reduction in revenue to local school districts and county road and bridge funds, all areas extremely important to Farm Bureau and its membership.

Judge Parker wrote: "The woodpecker makes no great or even necessary contribution to the ecological balance, his song is unremarkable and his plumage causes no heads to turn. However, he is on the endangered species list, and the law is specific about his protection."

Angelina County Farm Bureau President Hector Garcia contended, "The scales of justice are supposed to have two buckets to weigh the issues, but on this issue the scale only has one bucket, and the woodpecker is sitting in it."

Farm Bureau realized that it might be criticized for appearing to be taking a stand against the environment. But its actions were underscored by the words of Barry Ogletree, owner of the Cal-Texas Lumber Company. He said, "What a lot of people don't understand is by limiting the cutting in those areas where the woodpeckers are, you're virtually insuring the destruction of the exact bird you're trying to preserve.

"When cutting is limited, hardwood trees will grow in the areas where the woodpeckers are because that's the natural order of things in the forest. You have a woodpecker that's living in a pine tree. It's a dying pine tree at that. So it dies and nobody can go in there and cut anything so it's going to end up a hardwood forest. And the birds do not live on or in hardwood trees."

TFB President Bob Stallman pointed out in 1994 that many considered farmers to have a disregard for the environment, but he said that farmers in reality were the original environmentalists. He underscored the fact that farmers and ranchers live on the land with their families and drink the water there. He said, "They are not about to do anything that will harm their land. There have been cases where farmers have made mistakes, but as a whole, farmers and ranchers do an excellent job of taking care of the land."

No less a threat to private property, whether it be farm and ranch land or liquid assets, is an estate tax system that virtually confiscates the assets of each generation and prevents their passage reasonably intact to succeeding generations of descendants. The justification of an estate tax of any sort is the assumption that significant wealth must by nature have been obtained unfairly and thus should be captured and redistributed among the less fortunate, who are assumed to have also been less greedy. Most such schemes tend to be tempered by reality to the extent that their "recovery" thresholds begin at a point well above the level of the average estate. Rapid inflation and poor attention to adjustments, however, can have the effect of reducing that threshold to an intolerable level, especially when the bulk of the inheritance is land, which is historically overvalued in proportion to its productive capabilities.

In Texas, the result was an inheritance tax system that had become truly confiscatory by the 1970s. With Farm Bureau guidance, the Texas Legislature in 1976 revamped state law to peg the inheritance tax at the same level as the federal estate tax exemption, meaning that the state of Texas would only receive whatever would otherwise have gone to the federal government, but not a penny more.

Warren Newberry recalled, "Perhaps the most significant legislative program I have been involved with was the one that had to do with estate tax. As land values began to accelerate, a family would have a farm or ranch that would be worth easily $100,000 or $200,000 or more. The family might not have much money, but they had that land. If papa died, however, mama is left with the land, and she had to pay estate taxes. When mama died, the children had to pay it again. Sometimes they had to sell the land to pay the taxes.

"The Farm Bureau took the leadership changing that, and the new law had a great deal to do with keeping farm and ranch units together for the coming generations. It may have been our greatest accomplishment and not often understood or accepted as one of the greatest."

Two years later, again with Farm Bureau leadership, the legislature amended the Texas Constitution to allow the tax valuation of agricultural land according to its productivity rather than an arbitrary market price. That one change in taxation is considered by many sober observers to have done more than any other law to slow the conversion of farm and ranch land to development; by taxing land according to the "market" value of developed property nearby, prior law had virtually assured that legitimate farmers and ranchers bordered by developers sooner or later would be forced to sell for development as well, perpetuating an ever-expanding ripple of suburban sprawl. The new law provided a retroactive penalty for development, discouraging conversion both actively and passively, yet allowing landowners to make the final decision themselves, thus holding true to Farm Bureau policy on property rights.

A furrow irrigation system provided crops the critical water they needed to survive and produce.

And Farm Bureau was also in the lead when it came to protecting water resources, without which even the most desirable property may be of no practical use. Ironically, Texas had such strong private water rights for so long that the Texas Farm Bureau briefly found itself hamstrung when the tide began to turn in the 1980s.

As a "right of capture" state, Texas' water law has long paired water and land surface rights; whoever controls the surface controls the water beneath it. As a consequence, most Texas land valuations include the value of underlying water, and Farm Bureau policy staunchly supported its members' rights to "capture" the water under their land without government interference. That policy served the organization well until a series of dry years in the early 1980s prompted legislators to draw up a statewide "water plan" that included the specter of groundwater regulation.

Given the crisis atmosphere that drives so many governmental moves, the publicity that the state's water shortage had received, and the broad spectrum of interest groups that saw something to gain from a formal water plan, the success of some major piece of legislation appeared inevitable. What it would look like was the only real question.

Farm Bureau legislative personnel were caught between a tree and a stump. As a practical matter, many of the organization's members had already made a separate peace with groundwater regulation, even to the

A sprinkler irrigation system symbolizes Texas Farm Bureau's position to protect the landowner's ground water rights.

point of helping to organize local underground water districts. As long as they and their neighbors held the reins, the concept didn't scare them unduly; it was when control lay in the hands of outsiders that they got their backs up. Farm Bureau stood a good chance of influencing the developing water plan to assure that any groundwater regulatory program it included would be run locally. Without such a strong guiding hand, there was at least as good a chance that the plan would impose central controls.

As a frustrated Farm Bureau spokesman confided to a reporter during the thick of the fight, "I should be lobbying for local control, but our policy opposes groundwater control of any sort. If I do what I should do, I could lose my job, and they'd be right to fire me because I'm paid to uphold Farm Bureau policy. Unless and until the policy changes, I'm officially on the sidelines."

That change had to originate at the county level, where all TFB policy is spawned, then gain acceptance statewide. It did both, filtered in between through a sixteen-member special committee appointed by then-president S.M. True and chaired by TFB Vice President John Baker, a wheat, feed grains and beef cattle producer, from Bell county. The issue provoked some of the most active individual and county-level political activity within the Texas Farm Bureau up to that time. In the end, Farm Bureau played a strong and positive role in the Texas water plan, which addressed not only groundwater protection and conservation, but surface water allocation, reservoir construction, and pollution issues for years to come.

Since 1989, Texas Farm Bureau has been heavily involved in a crisis that arose when the Endangered Species Act was applied to the Edwards Aquifer. Lawsuits had been filed by the Sierra Club, challenging the right of surface owners—farmers and ranchers—to use the water under their land.

Joe Maley, TFB Public Affairs Director who provided staff leadership on the issue, pointed out, "Texas law recognized the principle of the "right to capture," and that means if there is water under my property and I can bring it to the surface, then I can capture it and use it as long as it benefits my family or my land. I cannot waste it, and I cannot contaminate it. But as long as it's for a beneficial purpose, I'm free to use it as I see fit. That is a very important issue to farmers and ranchers. They have to depend on that ground water sometimes for economic survival."

From 1989 to 1995, Texas Farm Bureau invested about a quarter of a million dollars, maintaining the "right of capture" for farmers and ranchers. And when the issue gained national implications, the American Farm Bureau shelled out more than a half million dollars in an effort to keep the Sierra Club from successfully bringing the Edwards Aquifer groundwater under federal control. As Maley explained, "It would be the first time the federal government has ever been involved in groundwater management. And that is something we do not want. That's something that has to be reserved to the states."

It's a crucial battle.

And it goes on.

TFB Executive Director Vernie Glasson said, "The fight could not have been waged without the courage and leadership of State Director Tommy Walker, his wife Ann and their Uvalde County Farm Bureau President Danny McFadin. These leaders, along with Uvalde County farmer Carl Muecker, Jr., put their names on the dotted line to get Farm Bureau and agriculture a seat at the table in the lawsuits."

The fact that Texas Farm Bureau leaders never compromised the group's principles despite the difficulty they faced in adapting to them only enhanced the credibility of TFB's position. Self-styled political sophisticates are conditioned to scoff at the idea that any organization has true "grassroots" direction, but anyone who watched Texas Farm Bureau in action during work on the Texas water plan in 1985 can testify that the term is honestly applicable to at least one group.

Texas Farm Bureau: The Members Always Rule

One of the most deeply held tenets of the Texas Farm Bureau philosophy is that its members must always exercise control, and that principle traces directly back to Walter Hammond. In Hammond's early days as TFB president, the organization had virtually no staff, so Hammond himself handled administrative duties. He was never really able to give up that control no matter how many staffers Farm Bureau eventually employed, because in his mind he represented the membership. He was the keeper of the gate.

It might be easy to attribute that attitude to a strong ego, but Hammond was sincere and consistent in his belief that grassroots farmers and ranchers, not bureaucrats — public or private — should control the destiny of agriculture. He was so sincere that he was willing to take on the entire federal Agricultural Adjustment Administration over it.

The battle was launched by Hammond's predecessor at Farm Bureau, Cliff Day. Following his departure from the Farm Bureau, Day rejoined AAA, where he and other government-oriented officials attempted to expand the role of county Triple-A committees beyond that of arbiters of federal subsidies. Their goal was to make themselves the primary voice in setting government agriculture policy. It was to be a top-down, rather than bottom-up program, run by men who could take a strong, centralized view of what farmers really "needed," and lobby for what they considered the farmers' best interests whether the farmers liked it or not.

Among the group's plans were commune-like land distribution schemes that would set a starved-out farmer down on ten or twenty acres of government-repossessed land where some previous farmer had already starved out and give the new man an opportunity to starve out yet again. It was an idea unabashedly imported from the Soviet Union, where similar practices were already turning that once-bountiful land into a net importer of basic foodstuffs.

The group's major roadblock was Farm Bureau, on both the state and national levels, and Texas was one of the biggest battlegrounds. The fight came to a head in the early 1940s, when Day's group hit on the tactic of beating the Texas Farm Bureau by joining it — or more to the point, by taking it over. Day had discovered a glitch in Farm Bureau's bylaws which allowed the outsiders to seat a slew of new directors. Once in place, it appeared they had the power to oust Hammond — TFB presidents were still elected by the board at that time, rather than by vote of the membership — and install a new leader sympathetic to their position.

The showdown came at a meeting for which no minutes and only sketchy recollections survive. What is clear, however, is that Walter Hammond stood the invaders down; when the vote was tallied, Hammond was still president of Texas Farm Bureau, and the takeover had failed. The renegade directors gradually either faded away or adopted Farm Bureau's positions. Describing Hammond decades later, longtime TFB field man Ed Cumbie of Bronte offered some insight into how the revolt might have been squelched.

"Walter Hammond," Cumbie remembered, "seemed to have a particular ability to work with people and to lead people, and he had the ability to sway people to his way of thinking." But Cumbie also described Hammond as an "open-minded" man who "was willing to listen to people and willing to negotiate." He recalled one of Hammond's many efforts to settle a dispute: "The only reason you and I can't agree," Hammond told an opponent, "is because we don't have the same set of facts."

"It was always his desire that people understand both sides of an issue," Cumbie explained, "and then they'd arrive at a common understanding."

As the organization grew and administrative loads increased, Hammond's unbending view of the professional staff's "place" at times tended to interfere with the flow of business. Eventually, it broke up an old partnership. The rift was between Hammond and his first professional organizer, Marvin Carter, and in hindsight it may have come about because of a rare instance in which Hammond overlooked his own advice to understand both sides of a dispute.

Marvin Carter by all accounts was intensely loyal to both Farm Bureau and Walter Hammond, and Hammond was a man who appreciated loyalty. The battle for farm-to-market road legislation had taken a heavy toll on the Farm Bureau president. On October 1, 1952, he was stricken with a grave artery condition while attending an AFBF board meeting in Chicago. For days, his life hung in the balance, and it would be three weeks before Hammond was released from the hospital and allowed to return home to Tye.

Vice President C. H. DeVaney was at his home in Coahoma when he received the stunning news. He immediately hurried to the state Farm Bureau headquarters in Waco to orient himself to responsibilities and duties he had just inherited, if only temporary. It was a big step—from relative obscurity to the leadership of the largest farm organization in the state and one of the most influential in the nation.

The *Texas Agriculture* reported: "It was no easy role to fill. DeVaney had stepped squarely into the shoes of the man whose inspirational leadership had made the Texas Farm Bureau a power to be reckoned with on both the state and national level. In a way, the convention of 1952 was a tribute to that leadership and that man. The president was absent, but his spirit permeated the assemblage. When he was praised in eloquent tribute, there was not a dry eye in the great assembly hall. Having been a close friend to the ailing leader for a long time, the vice-president felt Hammond's absence more acutely than most. That made the job more difficult, but easier, too, since it gave him the incentive and courage to carry on."

Hammond's health problems greatly concerned Marvin Carter, both for the sake of Farm Bureau and Hammond himself. Carter proposed that Hammond step down and turn the Farm Bureau leadership over to vice president DeVaney. The proposal, it can be safely said, was not to Walter Hammond's liking. He refused to step down, and indeed, would serve several more years before other health problems finally drove him from his life's work.

Hammond also turned on Carter, ordering his demotion, in effect, by splitting the state into two zones and offering Carter the choice of zones. Having been statewide organization director, Carter could not accept a post as a regional director, and he left Farm Bureau. The general assumption was that Hammond had punished his old friend for recommending his resignation, but Hammond's secretary, Creola Wobbe, saw the incident from close up and understood it somewhat differently.

The "thing that held priority in Mr. Hammond's thinking," Mrs. Wobbe recalled, "was that (Carter) had approached members of the board and discussed various things with them" during Hammond's illness. "I think they could have ironed out their other differences, "but not that one ..."

When J.H. West eventually replaced Hammond in 1958, much of the old one-man iron rule left Farm Bureau, but the concern for grassroots control which underpinned it did not. Texas Farm Bureau remained tightly regulated by written policy and bylaws, all of which originated at the county level and were adopted — or altered — only in open session by vote of locally chosen delegates.

West had long ago come to the conclusion that more of the problems that a farmer had to deal with lay beyond the fences of his farm than were found in the fields. He then dedicated a large portion of his life helping to solve those problems.

West understood agriculture. He began farming on the halves with his brother in the fertile coastal bend region of Texas, using mule-powered plows to turn the rows that grew his cotton. Within two years, the young men had accumulated sufficient capital to buy equipment of their own and rent land on the traditional third-and-fourth basis.

Harrell West early became acquainted with the unstable nature of farm prices. He observed that when crops were good and prices high, producers were able to pay their bills. The uncertainty of cotton prices, however, made both farming and merchandising disturbingly precarious. He later learned a great deal about what happened to cotton and other agricultural commodities when they left the farm. Through the years, he was therefore able to study the economics of the cotton industry, becoming one of the state's recognized authorities on all phases of the industry, from production to marketing to processing.

As a producer-delegate from Texas to the National Cotton Council, West pointed out that the cotton industry—from the grower on through retailers of the finished product—must cooperate in an intensive program of research and sales promotion, that new and better uses of cotton must be discovered and developed, that the surest and soundest way to prevent increasing surpluses of cotton—or any other commodity—was to create a greater demand for it.

Before assuming the presidency of Farm Bureau, Harrell West had served as secretary-treasurer of the organization for almost two decades. In fact, one of his primary duties during the early years was to garner, preserve and improve the meager financial resources of the bureau. In those eighteen years, he saw the Texas Farm Bureau grow from a small, weak group into a powerful, well-financed organization of more than seventy thousand members.

During the five years he served as president, the membership grew to 84,842. Three new departments were established: TFB's Research and Promotion Department (currently Research, Education and Policy Development), the Purchasing and Personnel Department and the Commodity Department.

Texas Farm Bureau had not forgotten how C. H. DeVaney had stepped in during Walter Hammond's illness in 1962 and served as head of the organization for four months until Hammond's return. He had definitely proved his mettle during the legislative session, handling Farm Bureau's legislative program with District 3 Director Loys D. Barbour of Iowa Park. The *Texas Agriculture* reported: "The record speaks for itself. Thirty bills favored and supported by TFBF were passed. Five important bills opposed by the Farm Bureau failed to pass...."

In 1962, DeVaney, who began his first farming operation on rented land almost three decades earlier, was elected to succeed Harrell West as president of Farm Bureau. He had joined the organization when it was still known as the Texas Agricultural Association and was only twenty-three years old when he cast his vote supporting the organization's affiliation with the American Farm Bureau.

DeVaney had a part in building the Howard County Farm Bureau into one of the strongest units in the state federation. The pride of his efforts was a brick migrant labor camp in Big Spring which was built with the help of local county and city governments. Howard County Farm Bureau also handled the contracting for Mexican Nationals.

The *Texas Agriculture* wrote of the organization's new president: "DeVaney is a native West Texan, and his good-natured, plain-spoken manner reflects his farm and ranch background in the wide open spaces...His presence is commanding...His inherent honesty and friendliness makes him a natural leader of men...."

DeVaney did indeed have a broad experience in leadership responsibilities with Farm Bureau. Early in his career on the board, he had served as chairman of the Labor, Rural Roads, Water and Legislative committees. He was perhaps best known, however, for his activities in helping to secure a workable animal health program for the state. He worked to organize the Texas Brucellosis Committee in 1952, serving as chairman when the group later became the Texas Animal Health Council, instrumental in securing the Brucellosis Control Program for the state.

He recalled, "During my presidency we got the idea that we could sell fresh beef to Europe. There was a demand for it. It looked like a natural because we would have a chain operation beginning when the calves are born, taking them on through the feeding, then the feedlots, and from that into the packing house, then overseas. But Europe didn't want our beef coming in over there. We carried one truck load—forty some odd head that we butchered and put in a container and carried it over to Germany. We sold it. We sold one to the army, then we sold one or two truck loads to chain stores overseas. But Europe was afraid we were going to take over their market, so they slapped an embargo of 100 percent of what it cost to get that beef to market. That kind of broke down our program a little bit."

Back when he was president of his county Farm Bureau, Devaney learned a good lesson. He saw first hand how difficult it was sometimes to unite on a single issue affecting farming. Farmers and ranchers might

The sale of beef to Europe seemed like a natural program for Texas Farm Bureau to adopt.

argue emotionally over an issue, but generally each of them had a legitimate and an honest point of view.

He said, "In Howard County, we decided to get rid of the Johnson grass out there. We made a deal with a firm that made a poison that you could put on it and kill it. We spent a lot of money out there. The farmers and the county cleaned up the barrow ditches and the roads, and so by gosh, there were some counties brought in a resolution that we eliminate Johnson grass all over Texas. So we got a bill introduced in the Texas Legislature to even make it illegal to plant Johnson grass.

"I was down at Austin lobbying. There was an old senator from over in East Texas, and he got his constituents all worked up over the bill. There were four hundred farmers in the House when they started debating the proposal to get rid of Johnson grass, and they were all mad. Why? Because they planted Johnson grass for hay. I could see why they were against outlawing Johnson grass. It's good hay, and they depended on it."

It was good to eliminate Johnson grass in Howard County. But leave it alone in East Texas. The solution to a single problem didn't always work everywhere.

Sidney Dean, named for the famed Civil War Confederate general who died at Shiloh, Albert Sidney Johnson, followed DeVaney to the president's chair of Texas Farm Bureau. Dean had been born and raised on a Louisiana family farm. He had entered the army during World War II, rising from lieutenant to lieutenant colonel while serving both as a detailed member of the War Department General Staff in Washington and on the Joint Staff in the Pacific Theater of Operations.

Dean raised small grain, corn, cotton and cattle in Victoria County, serving as a director of the Victoria County Farm Bureau for four years and Policy Execution chairman for two years. He became a grain sorghum and Brahman-Angus cattle operator and was named to the TFB Board of Directors in 1961. He was vice president of Farm Bureau before being elected to the organization's top position.

Texas Farm Bureau, historically, has always been involved in a wide range of national issues, even if they sometimes did not pertain directly to agriculture. For example, President Sidney Dean, in November of 1969, wrote Vice President Spiro T. Agnew when opposition to the war in Viet Nam began rioting in the streets:

"Voting delegates of the Texas Farm Bureau, representing 116,622 member-families of this voluntary farm and ranch organization...directed me by unanimous vote to convey to you the following message: As members of that silent majority of the United States citizens who normally refrain from demonstrations, revolutions, violence and general disruption of this most successful government and society the world has ever known, we do hereby vote to extend our congratulations and best wishes to the Vice President of the United States for his forthright opposition to and denunciation of the disruptive activities of a highly vocal and undisciplined minority in our society whose recent actions have undoubtedly given aid and comfort to the enemies of this nation...."

On the same day, Dean wrote President Richard Nixon, giving him a resolution passed by the Texas Farm Bureau Annual Convention:

"WHEREAS, we are engaged in conflict with an enemy in a foreign land...

"THEREFORE, be it resolved that voting delegates...do hereby affirm by unanimous vote the stated policies of the President of our United States of America in regard to bringing the Viet Nam conflict to an early and honorable end...."

Dean's letters brought notes of appreciation from President Nixon, as well as from both Texas Senators, John Tower and Ralph Yarborough, as well as Congressmen Kika de la Garza, Omar Burleson, George Bush, Bob Price, George Mahon and John Young.

In 1969, the 36th annual meeting of the Texas Farm Bureau, chaired by Dean, had a dramatic impact on the organization's structure. From the beginning, directors could serve indefinite terms. But Texas Farm Bureau changed that, setting a term limit in an effort to surface new leaders with fresh ideas that focused on different opportunities for TFB.

According to the new by-law amendment, each director would be elected for a term of two years, serving until his successor was named, and the director would be elected on a "staggered" basis. Odd-numbered districts elected directors in odd years, and even numbered districts in even years. Board members were eligible to serve for a maximum of three consecutive elected terms.

A second amendment called for the president of Texas Farm Bureau to be elected by the voting delegates present at the annual convention. The membership had grown larger. It was more educated and more involved in issues facing farmers and ranchers in rural Texas. It wanted a stronger voice on who should be leading TFB. This was, according to Warren Newberry, "a major, major change." The president was no longer the day-to-day administrator of Farm Bureau. As Newberry said, "The delegates took away the board's power to elect a president, but they then gave the board the power to employ an administrator, which kept a lot of power in the hands of the board."

President Dean asked Newberry who he thought could handle the new executive director's job.

Newberry told him, "Hire the very best person in the United States to do the job."

"I'm not sure who that is."

"Look," Newberry said, "you know everybody in the Farm Bureau nationwide. I think you'd look there first. Who is the best organization man you know in Farm Bureau."

"O.R. Long," Dean replied. "He's the director of field services for the American Farm Bureau. He's the best, but we can't hire him."

"Why not?"

"He's only a few months away from retirement."

Newberry shrugged and told Dean, "Well, I know you and your personality, and I believe you can hire anybody you set your mind to."

O.R. Long was intrigued with the thought of taking over as executive director for Texas Farm Bureau. Never a man to back off from a difficult chore, he accepted and moved to Texas. In Waco, Long set to work immediately applying his tremendous skills as an organizer and planner. His method was a simple one: "provide voluntary leaders an opportunity to get together and solve problems."

Newberry said, "O.R. Long had more influence and impact on the American Farm Bureau than probably anyone else. He did not mind staying in the background. If you were in charge of a program in a state, he would help you develop a program. He did not want nor get any credit. However, he was the prime mover. He had a vision of what the Farm Bureau should be like, and he helped state leaders move along, a step at a time, taking a bite they could chew. He did it by working with them."

With AFBF, he would travel into a state, visit with the president, the information director and organization director, which is about the only staff people most states had. After leaving, he wrote each, listing three or four things they could do to move their individual programs forward.

When he returned six months later, his suggestions were usually already in place and at work.

As Newberry said, "Long was a great teacher. He did it by example, and he was a patient person. He was very aggressive, but he was aggressive in a low profile manner."

Dan Pustejovsky, former state director and Hill County farmer, remembered the leadership schools O.R. Long conducted at Texas A&M University. He said, "Long handed out these tests. One of the questions said, 'We're interested in efficient meat production. There is a new kind of rabbit that has been developed. This rabbit will produce five times as much meat, half the amount of uniformity, etc.' Then the next question was, 'Do you want to raise this rabbit?'

"Everybody wanted to be more efficient. So we all said, sure, we wanted to raise the rabbit. The thing was rigged. We all got a zero on the test.

"Long told us, 'What you didn't know was that the rabbit had one bad fault. He was a carrier of Tularemia, which is rabbit fever. It infects people, and they die. He was teaching us to get all the facts before we made any decisions. He was teaching farmers and ranchers not to get sucked into something that doesn't work, and every farmer there, at one time or another, had known about peddlers selling them some snakeoil degreaser. Mr. Long just wanted us to know what we were doing before we did it."

During its development stages, the expanded program, under President J.T. Woodson, involved literally hundreds of top county Farm Bureau leaders. Under Long's supervision, meetings were held day and night all around the state. The sound and far-reaching expanded program of activities was the result of this grass roots approach to solving problems in Agriculture, and Long provided the organizational guidelines to make it work, permitting county leaders to perform their best.

Tommy Klemcke, serving as director of marketing, recalled, "O.R. Long was the ultimate administrator. He was constantly organizing and working with the staff to implement programs. He was kind of like a director on a movie set. He was always behind the scenes working, and he wanted everybody else out front. And that was kind of unusual for an administrator. But he wanted to build leaders."

His road to success was surrounding himself with the most talented people he could find. O.R. Long always said, "If you've got a bad leader, you don't try to push him under the rug, you put him out front and let the people decide whether or not they want to keep him or not."

J.T. Woodson stepped into Farm Bureau's presidency in 1970, just in time to guide the organization's expanded program, which focused on the establishment of special interest "commodity divisions." Woodson had been born on a farm in 1917, operating a gin in his hometown of Gober for fourteen years before deciding to make farming his life's work.

Woodson was elected a county Farm Bureau director in 1951, his first year in the organization. A year later, he became president, a position to which he was re-elected each year until he retired in 1956. Recognized for his leadership abilities and his knowledge of agriculture in general, Woodson was appointed to serve on the state resolutions committee for

The organization's expanded program established a commodity division to assist producers in marketing their produce.

three years. Always searching for ways and means of improving the Farm Bureau in his home county, Woodson founded the *"Fannin County Farm Bureau News,"* serving as editor. The experience he gained working so diligently for his home county gave him the knowledge he needed when the time came to take his leadership abilities to the state level.

Woodson served on the board of Southern Farm Bureau Casualty Insurance Company from 1963 to 1967, then again from 1971 to 1975. He was also president from 1972-1975. Woodson also served on the board of Southern Farm Bureau Life Insurance Company from 1963 to 1967 and from 1971 to 1975.

O.R. Long continued working with Woodson as executive director for only twenty-one months. The important expanded program was brought to reality. Membership increased by more than 10,000. And financial stability was achieved. He was finally ready for retirement, ready to pass the torch to another.

Warren Newberry, who had been serving as director of the Field Services Division of the American Farm Bureau, succeeded Long as executive director. Newberry was on hand when the American Agriculture Movement came rolling into Waco to attack the Texas Farm Bureau with a tractorcade. He wasn't concerned. He said, "We've always had this very radical element in agriculture." The farmers tried to unionize and not sell any produce at all until they were able to get a good price. "And we'll quit farming until we get it," they said.

But that was tough to do. Many of them were already going broke. Often they were third-generation farmers, and many times they were on the threshold of taking their fathers and grandfathers down with them because they had borrowed too much money to raise a crop that would not sell.

Newberry said, "They were sick of it. And they got sympathy from the people who weren't down yet but thought there was a pretty good chance they would be unless farm prices turned around."

The movement spread suddenly and rapidly. The radical ideas caught on and exploded across the country.

Newberry received a phone call one afternoon. The message was ominous. It said, "They're gathering, and they're on their way."

The tractorcade rolled to a stop in front of the Texas Farm Bureau office. Newberry said, "There must have been a hundred tractors, and there were no old clunkers among them. Everybody was driving the bigger, newer tractor, which was interesting because they all said they were going broke."

The angry farmers surrounded the building, saying they wanted Texas Farm Bureau to be on their side and take a position that guaranteed prices for agriculture products. They also demanded to know why Farm Bureau had not already taken that position. They were mad. They were young. They were ready to revolt. They built a fire out in the middle of the grounds and burned their Farm Bureau membership cards and swore they were burning their insurance policies as well.

But Newberry recalled, "They have to come over to the county office to cancel their insurance. And we didn't have a single cancellation."

When Carrol G. Chaloupka, a livestock and grain producer from Dalhart was elected president of TFB in 1974, he brought with him a solid background of being able to fight for Farm Bureau programs. He had been a pilot of B-17s and B-29s during World War II. Active in the reserves, he was called up for service again during the Korean conflict, retiring from the Air Force Reserves in 1972 with the rank of lieutenant colonel.

Chaloupka had come to the Plains during the hard, unforgiving 1930s, borrowing a hundred dollars to buy cattle. In time, he was producing wheat, grain sorghum and livestock on a 10,200 acre farming and ranching operation. Chaloupka had been active in Farm Bureau for years, serving three years as president of the Hartley County Farm Bureau. In 1969, he was elected TFB state director for District 1, then chosen secretary of TFB the following year. In 1972, his steady rise to the top continued with his election as vice president of Texas Farm Bureau.

The Fort Worth *Star Telegram* described Chaloupka as a "hybrid product of his environment, a firm believer in the free enterprise system, who doesn't believe sundown means quitting time...a hard working, sincere and conservative Czech."

Chaloupka's farming ability was translated to his role as president. The organization noted his astute management ability that had enabled him to make money on cattle in twenty-eight out of thirty years, and his cagey gaming with the markets to increase his profits on wheat.

In 1982, there were even speculations that Chaloupka would be a candidate on the Republican ticket for Texas Commissioner of Agriculture. But he declined the opportunity, saying, "I am totally committed to my family, my farming and ranching operation, and to the Texas Farm Bureau, and therefore could not even entertain the thought of a political race."

Warren Newberry, executive director at the time, remembered, "Carrol was a very aggressive, strong-willed leader. He had ideas of his own and had the capability of getting county people's confidence. County people were fiercely loyal to Carrol. He had that ability to go out and talk to them and win their confidence and their support. He pushed hard for many things, trying to make things happen, and I admired him for that."

In 1982, S. M. True, a Plainview cotton, grain and cattle producer, ascended to the presidency of Farm Bureau. He had held every office in the Hale County Farm Bureau and was first elected to the TFB board of directors in 1975.

Of S.M. True, Warren Newberry said, "He brought something to the Farm Bureau movement that's unique, and that's called stability, a first for the Texas Farm Bureau. TFB is one of the most—is the most—volatile state farm bureau. But Mr. True brought stability to Texas.

"He was so level-headed and great under pressure. When things get tough is when you're really tested. I've seen S.M. True under all kinds of very difficult circumstances, but I've never known him to panic or lose control. It's his personality. He was a night fighter pilot in World War II, flying off the decks of aircraft carriers. Even in business, the tougher it is, the calmer he is. He has more empathy and respect for the professional workers in the Farm Bureau than anyone I have every worked with."

Vernie Glasson, who had served as both assistant director of national affairs and field services director for the southern region of the American Farm Bureau Federation, became executive director when Newberry retired. In 1984, he had been brought back to Texas Texas Farm Bureau by True to direct the public policy/governmental affairs program of TFB. And he remembered: "S.M. True epitomizes everything that I have ever expected or hoped for in an agricultural leader. He's a man's man. When he told you he was going to do something, he did it. He was always caring."

S. M. True worked hard for the membership, believing that the success of Texas Farm Bureau was directly related to the success of the members.

True was on the board of Southern Farm Bureau Casualty Insurance Company from 1978 to 1994, serving as president and chairman of the board from 1990 to 1994. He was a member of the Southern Farm Bureau Life Insurance Company board from 1980 to 1994.

S.M. True had the distinction of chairing a special called convention of TFB voting delegates. In Farm Bureau's six-decade history it has been necessary only twice to convene a special meeting of voting delegates for an emergency bylaw or policy amendment.

The first such gathering, gaveled to order by president J.T. Woodson, was in March of 1971 at the Baker Hotel in Dallas. The situation that precipitated the previously unheard-of called session was a proposal to expand Farm Bureau services into commodity programs, marketing, tax and estate planning and public relations. The plan called for a hefty dues increase to pay the freight. Whether or not the matter deserved a precedent-setting special meeting remained a bone of contention throughout the day-long meeting; many delegates — mostly opponents of the expansion plan — insisted it could as easily have waited until Farm Bureau's regular fall convention. They were outvoted, however.

The plan was the result of several months' work by a special thirteen-member study committee. What made the proposal unique was that each element of the package had originated with one or more county farm bureaus — not among staffers seeking to expand their own programs or even with TFB's board of directors. In fact, some of the plan's most persistent opponents sat on the board.

In keeping with the organization's passion for grassroots control, voting delegates at the previous meeting had bypassed the board. They picked the committee, outlined the assignment and ordered the thirteen members to report back directly to the delegates; the TFB board would be kept apprised of developments, but not invited to rule yea or nay on the particulars. As with its origin, the meeting epitomized democracy in its purest form. A freewheeling affair, it was conducted in proper parliamentary fashion and with strict attention to neutrality on the dais. But sometimes exasperation got the better part of propriety.

The commodity and marketing proposals, which were somewhat vague and tentative by necessity, passed with little dissent. Opponents mostly voiced concern about the costs. In particular, producers whose commodities did not lend themselves to marketing programs were reluctant to help finance plans strictly for the benefit of others. Supporters were at a loss for viable cost figures because so much depended on the details, and the plans as they stood were only skeletal.

Expansion of TFB's education and research programs likewise passed with little trouble, as did plans to offer limited income tax and computerized recordkeeping services for members. Also approved was an increase in field staff to help generate more membership participation in legislative affairs. The idea in the latter case was to involve more individual farmers and ranchers in hearing testimony as well as providing direct contact with legislators. The measure had the backing of the professional government affairs staff, who were well aware that when it came to a crunch, faces from back home had more impact on a lawmaker than those of paid lobbyists.

The first proposal that ran into serious opposition was a plan to raise agriculture's visibility through a concerted public relations campaign; TFB would buy television and radio time and air professionally produced commercials aimed at the city housewife, who was at that time under considerable encouragement from the popular press to complain about the price of food.

"We are fast becoming a minority," explained committee member Jerry Salley.

Hays County producer Frank Watson agreed. "The price of meat goes up and (the housewife) goes to Mr. Nixon and he responds right quick with programs that reduce the price of beef. The only way we have for telling what meat is worth is to get on the TV by paying for it, not asking for free time." Otherwise, "they will convert your words the way they want it."

Former board member Loys Barbour, of Wichita County, spoke for many of the opponents when he quoted Roger Fleming, American Farm Bureau Federation secretary, who gave assurance that agriculture's public image was still good. Public polling, Fleming had said, found that "the public has a high regard for the farmer and appreciation for him for the role he plays in the world economy."

Other opponents worried about the plan's reliance on broadcast rather than print media, and still others contended that granting an advertising contract to one television or radio station in a given market would jeopardize what free publicity Farm Bureau might be getting from competing outlets.

What may have influenced the opposition as much as anything, however, was that the public relations campaign proposal was the only one that came complete with a dollar figure for costs. Money concerns seemed to drive opponents most of the day. Funding was such a powerful driving force, in fact, that late in the day it threatened to undo all that the delegates had done up to that point. After approving essentially everything that the committee had proposed, and maintaining a sense of humor despite moments of contention (TFB Vice President Mike Frost at one point counted votes by asking delegates in favor of one proposal to "stand up and stretch"), the same delegates voted down the proposal to raise dues enough to pay for the new programs.

"The board didn't approve of this," argued Loys Barbour in opposition to the dues increase, to which Frank Watson of Hays County retorted that "we ... as members are the grassroots of this outfit. We make our recommendations first, before the board tells us what to do. We tell the board what to do; we don't have the board telling us."

The original dues increase proposal called for hiking annual membership from five dollars to nine dollars, of which a dollar fifty had already been approved and earmarked for the advertising campaign. Efforts to increase the raise to ten dollars and to trim it to seven dollars and fifty cents had both failed when the original proposal finally came up for a vote. Because it would be a bylaws change — an example of how tightly Farm Bureau ran its ship — passage would require a two-thirds majority, or 456 votes given the number of delegates on hand. It managed only 362.

That left supporters in a quandary, and not at all happy about the state of affairs. "Where we stand," committee chairman Charles Allison reminded the delegates, "is that we have a program but no finances to put it into action. If we do not provide finances for this, it cannot be put into action unless somebody has awfully good credit."

Texas Farm Bureau's convention in Dallas promoted the benefits of the expanded program.

John Allen of Lubbock chastised those who voted against the dues hike. "I am thinking," he told the group, "about these reporters writing ... about everybody voting for the expanded program, and right at the last, they voted no money for it. It's ridiculous. I was against the original plan, but if we voted to spend it, let's get it up and let's don't haggle."

Others took up the challenge: "Most of you are going to spend $15 for a room tonight," admonished Ray Carnes of Uvalde, "and you're sitting up here arguing about a dollar fifty or six dollars and fifty cents. This is so stupid."

"Let's don't fertilize this field like we did today and then refuse to let it rain on it," insisted Calvin Reuter, Bosque County.

After a parade of such witnesses, enough delegates were properly chastened to compromise on a seven dollars dues increase and pass it with nearly forty votes to spare. Texas Farm Bureau's first special called meeting ended with locally-chosen delegates adopting essentially all of a slate of locally-proposed programs, and voting to pay for them with a locally-initiated dues hike.

Farm Bureau's unquestioned fealty to its members' direction is nowhere better exemplified than in the efforts of President S.M. True and TFB in the "Jim Hightower Story." That was a case in which the Texas Department of Agriculture was captured by a political machine whose philosophies, policies and actions were so abhorrent to so many Farm Bureau families that the organization's members voted overwhelmingly to reverse three of their most cherished and time-tested policies in hopes of recapturing "their" TDA.

Years later, Texas Farm Bureau Executive Director Vernie Glasson issued Jim Hightower a heartfelt — if thoroughly backhanded — "thank you." Hightower, Glasson explained, "did more for the Farm Bureau than maybe any friend we have ever had in the position, because he brought us together as an organization. Plus, he brought agriculture and Texas together in opposition to him.

"Whether we had argued over price supports before, over brucellosis, or whatever the issue happened to be, Jim Hightower kind of became a 'cause celebre' for all of agriculture and especially Farm Bureau."

Hightower got his political start as an aide to a liberal's liberal, former U.S. Senator Ralph Yarborough. When Lloyd Bensten captured Yarborough's seat, Hightower drifted back to Texas, settling in at the equally liberal *Texas Observer*. As editor of the *Observer*, he spent several years perfecting a "populist" shtick, goring conservatives of both parties — "Texas Republicrats", he called them — and building a name for himself in the left wing of the Democratic party.

Hightower made a failed bid for the party's nomination to the Railroad Commission, then turned his attention to another elective office, that of Texas Agriculture Commissioner.

It was perfect for Jim Hightower, and it was semi-vacant. After years in the office, John White moved to Washington D.C., to serve in the U.S.

Department of Agriculture under Jimmy Carter's Administration. White became Chairman of the party's national organization in the late 1970s. The ag post was awarded to long-time party functionary Reagan Brown, whose trove of anecdotes and one-liners made him a popular speaker on the agricultural convention and meeting circuit.

Texas was still very much a one-party state, so there was seldom if ever any serious Republican competition in the race, and unwritten protocol decreed that the party's anointed candidate would not be challenged in the Democratic primary. That combination had made the ag commissioner post a safe one for years, but it also made it a soft target for someone willing to buck the fading party hierarchy. A renegade like Jim Hightower. The darkest of dark horses when he entered the 1982 commissioner's race, Hightower was challenging a genial and popular incumbent, albeit a short-timer, for a post that virtually no one outside of agriculture cared a whit about.

Then Brown stumbled. The irrepressible jokester made the unpardonable mistake of telling a racially-tinged story in the presence of reporters. To that point Hightower had not had an issue that would gain him more support among the general public than it scared away among agricultural people, the only ones likely to cast a deliberate vote in the race. The "evils" of technology, and depictions of Ma and Pa Kettle as "family farmers" fighting off greedy "agribusiness" left voters yawning. Painting himself as a crusading reformer out to unseat a racial bigot, however, Hightower captured the Democratic nomination and later rolled over token Republican opposition in the fall.

Thus, Texas found itself with an agriculture commissioner who was openly hostile to production agriculture. Hightower's constituency was the Ralph Naderite "consumer" movement on one hand and Cesar Chavez' United Farmworkers Union on the other. He could be seen ranting against "tainted" apples during the Alar hoax, and he occasionally skipped a meal to demonstrate solidarity with Chavez during the activist's periodic fund-raising "fasts."

Despite the highly politicized nature of the commissioner's post, the Texas Department of Agriculture itself had long operated as an efficient service agency. Weights and measures, seed, feed and fertilizer inspections, livestock and grain market news and commodity promotions were its stock in trade, along with a common-sense pesticide regulation program. There was an invisible line between the political work of the agriculture commissioner and the practical work of the professionals he nominally headed.

When Jim Hightower took office, that line was erased. Longtime department heads and experienced staffers were jettisoned to make way for Hightower's personal promotions machine, and funds appropriated for traditional TDA programs went instead to build a high-tech public relations department whose primary purpose was to keep Hightower's name and face before the public in what everyone accepted as the buildup to an eventual run at higher office. Glossy newsletters and a veritable blizzard of press releases went hand in hand with slick radio and television spots.

Jim Hightower was everywhere: chewing on a grass stem while doffing his store-creased hat and mopping the sweat from a furrowed brow as he pondered the problems of a beleaguered farmer; occupying the best camera angle at the groundbreaking for an agricultural enterprise that someone else developed and funded; or pounding a podium somewhere, condemning conservative agriculture policies in the name of Texas farmers and ranchers who did not necessarily agree with him.

But it wasn't the egregious self-promotion — even at taxpayer expense — that soured Farm Bureau on Hightower. It was his policies. Hightower came into office with a draconian pesticide regulatory program already mapped out with the help of environmental activist groups. He backed consumer and farmworker boycotts of agricultural products, even to praising a European Community ban on beef byproducts because of this country's use of growth hormones.

And Hightower was one of the architects of a socialistic "alternative" farm bill during the mid-1980s that sought rigid production controls on all agricultural commodities. The hoopla surrounding that plan included a massive "convention" in the Midwest, where a carefully choreographed selection of "random" speakers had gullible media types convinced that the entire agriculture industry favored such a radical shift in policy. Some legitimate producers who had been talked into participating felt sheepish about it later; they began to sense the real drift of the program, they explained, when they saw that the organizers' handouts were rife with depictions of America as a global scourge, and insistence that capitalism and western technology were to blame for Third World poverty.

Texas Farm Bureau fought most of Hightower's initiatives in the legislature and was able to temper much of his ambitious anti-pesticide program. His ability to commit mischief in league with other state and even federal regulatory agencies made him a constant threat to agricultural producers, however, and TFB made Hightower's defeat a priority.

"Farm Bureau decided," Vernie Glasson explained, "that we had a problem and that we needed an advocate to speak for agriculture ... rather than an opponent trying to gut us every time we came upon a crucial issue. We became, in essence, a laughingstock ... having a commissioner of agriculture like we did in the state of Texas."

It took until 1989 for a plan developed by True to begin to jell. With his quick and unusually vicious tongue, Hightower was a media darling. That and four years of taxpayer-funded self-promotion allowed him to walk away with the 1986 election. Faced with the prospect of Hightower doing it again in 1990, Farm Bureau was ready to seize on any straw, and the straw that presented itself came in the form of the Sunset Commission.

By law, all Texas governmental agencies must undergo "sunset" review periodically; in the process, they are subject to anything from mild revamping to outright elimination. It was TDA's turn in 1989, and TFB saw

Top: Migrant farm workers unionized under Cesar Chavez and supported the candidacy of Jim Hightower.
Bottom: The black eagle was known throughout the entire country as the symbol of the United Farmworkers Union.

in the sunset process a chance to deal with Jim Hightower.

Farm Bureau's first approach demonstrated how desperate the situation had become: after decades of staunchly backing the election of the Texas agriculture commissioner, TFB opted to pursue a change to an appointive position. "We felt our problem had grown that severe," explained Glasson, "and we felt that, frankly, with the urbanization of Texas, we might not be able to change it in the political arena."

One presumed ace in the hole was then-Gov. Bill Clements, the state's first Republican governor since Reconstruction. In his highly partisan attacks, Hightower had taken to describing Clements as "bi-ignorant," so there was not a shred of doubt that the Governor would send the acerbic ag commissioner packing within minutes of getting the go-ahead.

There was just one problem: Farm Bureau's official policy still backed an elected commissioner. When the organization had last met, Hightower was making loud noises about giving up his post in 1990 to challenge U.S. Sen. Phil Gramm, so there had been no thought to changing policy. It was only later that Hightower backed away from a clash with Gramm and announced another run for ag commissioner. If TFB waited until its next convention to address a policy change, there would be no point in bothering because the sunset review would have been over and done for months.

So, in March of 1989, Texas Farm Bureau called its second special meeting chaired by President S.M. True. The agenda was short and to the point: did locally-chosen delegates favor an appointed or elected agriculture commissioner? By a sizeable margin, the delegates voted to break with decades of tradition and back an appointed commissioner. Farm Bureau's grass roots had spoken. It would all be for naught, however, in large measure because the Democrat-dominated legislature could not bring itself to turn such a valuable plum over to a Republican governor. As the legislative session drew to a close, the bill reauthorizing TDA made few if any significant changes.

Clements, an early backer of Farm Bureau's position, met with a TFB delegation during the autumn of 1988 in the governor's mansion. He looked S.M. True in the eye and said, "S.M., we have to have Farm Bureau's support." During the meeting, Clements indicated that he would veto any measure—including Sunset legislation—that failed to include an appointed agriculture commissioner.

At the time, Texas Farm Bureau had a policy position that supported the elected Commissioner of Agriculture. The conversation with Clements, however, made TFB re-assess its position and have an unprecedented Special Called meeting of the voting delegates in March of 1989. The only issue on the agenda was whether or not TFB would change its position and support an appointed Ag Commissioner.

Delegates voted to let Governor Clements know they wanted a change in the makeup of the leadership of the Department of Agriculture, a position that allowed Texas Farm Bureau to support an appointed commissioner.

Joe Maley recalled, "After we went out on a limb and had the special delegate session, Governor Clements changed his mind and signed the re-authorization or the Sunset legislation. It left the elected commissioner intact. And we were the ones that were hung out on that one."

Most of those close to the situation were convinced that Clements had double crossed Mr. True and the Farm Bureau leaders who trusted him. Regardless, Clements's vow to TFB evaporated. He claimed that a veto would eliminate TDA, and in the process cost agriculture its inspection, marketing and statistical services as well as the pesticide regulatory program. Given TDA's varied pesticide responsibilities under federal laws, he added, Texas producers could be barred from purchasing or applying restricted-use products as long as the regulatory program was shut down. Whatever the Governor's real motivation for changing his mind, he said that Jim Hightower would still be agriculture commissioner, because his was an elective post independent of TDA. A veto, then, would hurt everyone but Hightower, or so Clements said as an excuse to S.M. True.

That left only the electoral process itself for relief, and Farm Bureau had been beaten there before. So they tried something bold, feeling out the Texas Rangers' legendary pitcher Nolan Ryan as an opponent for Hightower. The only way to beat Hightower's high public visibility, went the reasoning, was to counter with someone whose visibility was even higher.

"He seriously looked at it," Glasson remembered, but Ryan eventually decided he was not quite ready to retire from baseball. Ryan had been a long-time member of Texas Farm Bureau. He was a rancher in both Alvin and in South Texas, raising commercial cattle. Presently he has in excess of two thousand mother cows on the two ranches he operates. But,

President S.M. True led the successful fight to change leadership in the Texas Department of Agriculture.

in 1989, he still had great moments in a great career ahead of him. He registered his 5,000th strikeout and put his seventh no-hitter on the scoreboard before retiring from the Texas Rangers in 1993.

"Nolan Ryan was, and is, a dedicated friend to Agriculture and Texas Farm Bureau," said Glasson.

Having been bold, Farm Bureau decided to be even bolder, and field a whole slate of candidates to muddy Hightower's water.

"Our goal was to have thirteen candidates, one per Farm Bureau district, running as Democrats, and we were going to fund their activities." There was no hope of actually defeating Hightower in the primary with such a tactic, "but we could send a message to Texans that we were serious, that we had a problem."

In the end, only a half-dozen Farm Bureau candidates actually filed, and Hightower quickly dubbed them "the six-pack." Comprising the candidates were Neal Burnett, Clyde Chandler, Dan Krenek, Paul McDaniel, Dan Pustejovsky and John Earl Smith, all running on the slogan, "Anyone But Hightower." As their brochure said, "We asked Jim Hightower to stand his qualifications before us. But Jim had none."

"I'm glad there weren't seven," Glasson joked later, "or he could have called them the seven dwarves."

As expected, Hightower won his primary handily, but five of the six TFB candidates captured their own counties, and the slate overall took a sizeable vote. That count, added to the votes cast in the Republican primary, suddenly gave Farm Bureau renewed hope that Hightower could be beaten. Their new weapon would be the Republican victor, Rep. Rick Perry.

An agriculture producer himself, the Haskell native was a match for Hightower's speechifying ability, and with his clean-cut good looks and confident manner, he posed a striking contrast to the antic, diminutive Hightower, described by one newsman as resembling "a Chihuahua on amphetamines."

The Perry campaign benefitted from what was actually the first and second of the three major policy changes Hightower brought about at Farm Bureau. Until the fight with Hightower, Farm Bureau had scrupulously avoided endorsing political candidates, and in the same vein, there had been nothing resembling a political action committee related to Texas Farm Bureau. That all changed in 1987, following the 1986 ag commissioner election in which TFB's policy had forced it to stand aside and bite its collective tongue while Hightower walked away with the vote.

Until then, Glasson said, "we just simply didn't want to get our hands dirty ... but we were one of the only groups, if not the only major player in Texas, that didn't have a political action committee." The idea had come up before, because many of agriculture's legislative friends were not getting much backing from the industry. Until Hightower, however, it didn't go anywhere. Another catch, Glasson discovered, was that most Farm Bureau delegates didn't understand what a PAC was all about, and their natural tendency was to vote against anything they didn't understand.

"So we spent 1987 working on a prototype for AGFUND, trying to put it on paper to show exactly how it would work ... and how Farm Bureau members themselves would control it."

The issue was debated on the convention floor. And, as Joe Maley recalled, the delegates realized that the whole purpose behind a political action committee or a special purpose committee was to allow a group of people with a common interest to pool their resources and be more effective, especially on a statewide level. That was what AGFUND would allow Texas Farm Bureau to do.

One by one TFB leaders marched to the podium to educate delegates on the benefits, the value, of AGFUND in an effort to secure the necessary votes to pass the proposal. The leaders included John Baker of Bell County, Steve Bearden of Cameron County, David Murrell of Chambers County, Jim Smith of Jefferson County, Bob Turner of Coleman County, Neal Burnett of Hale County, Don Smith of Hopkins County and a long line of others.

The proposal passed "overwhelmingly," and has since recorded better than an 85 percent success rate despite Farm Bureau's somewhat maverick practice of backing longshots and challengers as well as the "safe" incumbents most PACs rely on. The PAC collects its funds a dollar at a time, allowing voluntary contributions to be added to annual dues. Glasson said roughly fifty-five percent of membership renewals include the AGFUND contribution, and the program is working as it was originally designed. And Glasson should know, as he was the principal architect of AGFUND—or officially the Texas Farm Bureau Friends of Agriculture Fund, Inc.

The new fund played a role in Perry's campaign, as did Farm Bureau's decision to allow endorsements. But along with those benefits came a lot of individual sweat and labor.

"We had all kinds of fundraisers," Glasson recalled, "from pancake breakfasts to rodeos ... so that he could get on television in the urban areas. We had farmers who were pulling old abandoned cotton trailers out of their junk piles, parking them on the side of the highway and putting a big 4x8 'Rick Perry For Agriculture Commissioner' on the sides of interstates throughout Texas." Other members sent vote requests to all the urbanites on their Christmas card lists.

Farm Bureau also sponsored a "pickup rally" led by True that converged from all over the state and ended up at the Capitol, where a "baloney lunch" symbolized the fact that farmers and ranchers had experienced enough "Hightower baloney."

"The bottom line," Glasson said, "is that we had a unified effort underway in Texas among agriculture, and ultimately the urban public said these farmers and ranchers must be serious; they must really have a problem. It was just an excellent grassroots campaign, the likes of which I've not witnessed before or since, and it was really a good feeling.

"Mr. True, who rallied all of Texas agricultural forces together in the fight, made the statement as President of Farm Bureau many times that 'We're going to change the leadership of TDA.' He never said 'we're going to beat Jim Hightower or we're going to elect Rick Perry.' He said, 'We're going to change the leadership of TDA.' And we did."

Such a strong unity of agriculture became a significant factor in the political life of the state, as well as in Farm Bureau itself.

It helped that Hightower became one of his own worst enemies. Displaying the arrogance bred of overconfidence, he had pulled such shenanigans as hiring lobbyists to defend his job before the legislature, and those stories finally came out as the election neared. Using public funds to lobby on behalf of a state agency is highly illegal, but Hightower had well-known lobbyists working the halls and other operatives keeping him posted on key votes using then-rare and quite costly cellular telephones.

Hightower was also caught using agency credit cards to pay for travel, lodging and meals in connection with his campaigns for top-level Democrats in California and New York. Some of the expenditures were truly profligate. He and his lieutenants always had excuses for the transgressions, and the Travis County district attorney was loathe to inconvenience a fellow party leader, so Hightower himself escaped prosecution, but the public remembered.

As a final nail in the coffin, his chief deputy and two long-time political operatives were indicted and eventually convicted of using TDA's authority over feed and seed inspection to "shake down" ag supply dealers for contributions to Hightower's campaign. As always, Hightower dismissed the prosecutions as "politically motivated," and, as always, he himself slipped the noose.

But when the votes were tallied after the 1990 election, Jim Hightower was no longer Texas agriculture commissioner, and Texas Farm Bureau had been vindicated, not to mention successful.

"People were laughing at us," Glasson said of the failed effort to oust Hightower in favor of an appointed commissioner, and for fielding the "six-pack" of candidates in the 1990 Democratic primary. "They don't laugh at us any more."

PAST PRESIDENTS

J. Walter Hammond, president, 1939-1957

J. H. West, president, 1957-1962

C. H. DeVaney, president, 1962-1967

Sidney Dean, president, 1967-1970

J. T. Woodson, president, 1970-1974

Carrol Chaloupka, president, 1974-1982

S. M. True, president, 1982-1993

Texas Farm Bureau: Increasing the Benefits

The root of Texas Farm Bureau's early success was in its governmental relations, and it cemented its membership with insurance offerings. But by the 1960s, TFB was looking for other ways to provide services. Bill Wedemeyer recalled some farsighted leaders remarking that "if we are to build a stable membership there must be something for the member that the non-member cannot have."

The organization found part of the answer in cooperative purchasing, using its size and buying clout to provide goods and services at prices that members couldn't obtain elsewhere. As it had with the various insurance programs, TFB initiated a separate "affiliated" company, the Texas Agricultural Service Company, or TASCO, to operate the program under the capable management of A.R. Grauer.

In 1965, under C.H. DeVaney's presidency, Texas Farm Bureau began marketing the American Farm Bureau Federation's own "Safemark" brand of tires and batteries. The TASCO program initially involved the sale of premium quality car, truck and tractor tires and tubes. And, as Grauer pointed out, the new service-to-member tire program would mean a substantial operational cost savings to Farm Bureau members. He explained, "Members will be able to purchase tires that, for example, carry a suggested retail price of $36.25 for $17.00."

Grauer was proud of the fact that the full four-ply, top-grade Nylon construction was, pound for pound, stronger than steel. The tires had built-in features that not only lengthened the life of the tire through greater heat dissipation, but provided a safe, more comfortable ride by reducing sway and tire distortions. The magna-curve tread was designed to give added gas mileage by reducing wind resistance and friction between the rubber and the highway. The tires were, Grauer said, the kind of quality that could make a difference in the day-to-day living of Texas rural families.

In 1988 the program began operating under its own name, TASCO, no longer tied to the American Farm Bureau service corporation. TASCO is still going strong, but it got off to something of a rocky start in parts of Texas. Ed Cumbie remembered that he and the relative handful

TASCO was a TFB program that gave members an opportunity to buy good tires at inexpensive prices.

of fellow field men employed at the time were given the responsibility of signing up TASCO/Safemark dealers in their respective territories, of riding herd on those dealers, and of handling the adjustments that are inevitable with tires and batteries. That was all in addition to their normal duties of advising county offices, as well as overseeing membership drives. In Cumbie's case, it had to be done across a district stretching 450 miles east to west and 300 miles north to south.

"It was quite an interesting experience," Cumbie recalled with a heaping share of understatement. "You ran into lots of unusual situations" trying to learn the tire and battery business while engaged in it, in the process trying to stay at least even with people who'd been in the business all their lives and who could devote their full attention to it.

"One of the first people I signed up in the TASCO program," he remembered vividly, "was the largest independent tire dealer in Texas at that time." It looked like quite a coup, hitting such a big lick for TASCO right off the bat. With that dealer's volume, not only should Safemark sales have expanded rapidly, but non-members who saw the deals their neighbors were getting could be expected to pony up the modest Farm Bureau membership fee for a chance to participate in the savings. The dealer's large clientele looked like fertile ground for recruitment.

For some reason, however, none of that happened. Cumbie did some checking and "it became obvious pretty quick that the dealer was opposed to our program," that "his primary reason for wanting in the program was to try to suppress it. If he could get all his stores signed into the program in the various counties, he could — he thought — do what he wanted to with our tires."

For a while it worked. The dealer would put Safemark tires "back in a dark corner somewhere, and when one of our Farm Bureau members came in, he would say, 'Well, you don't want that tire. It's not the quality of this tire that I have here, so let me sell you my tire.'" Once those tactics were exposed, he did not remain a TASCO/Safemark dealer.

That wasn't the only such case. "A lot of people tried to abuse our program and to criticize us over the years," Cumbie said. But TFB ended up using the legitimate criticism to strengthen its TASCO effort, and it relied on loyal Farm Bureau members to help sniff out and expose the abuses.

TASCO expanded into tillage tools and vehicle accessories in the 1970s, and might have carried the expansion to greater lengths had it not been for some sage advice Farm Bureau received years earlier. Cumbie recounted an incident from the 1950s, when the organization was first thinking of branching out beyond insurance. An Illinois Farm Bureau representative addressed a meeting of county presidents to explain the Illinois program. "If you are being adequately serviced from any other source," Cumbie quoted him as telling his Texas audience, "then there's no need for Farm Bureau to get into the field."

That has been a driving philosophy behind TASCO and other Texas Farm Bureau service programs as well. From the inception of the purchasing program, Cumbie explained, "We were not getting into it just to be going into business, but trying to provide a service to our members. If we can't provide the service to an advantage, then why start it?"

The key word in the Illinoisan's advice was "adequate," Cumbie explained, and that is a subjective concept applying to both service and price. Over the ensuing years, TFB has expanded its "Service-to-Member" programs into a host of areas where it considered either price or service to be inadequate. In the health field, which kicked off Farm Bureau's services in the first place, TFB has added rural health services, "Coast to Coast Vision Plan," dental insurance and mail order prescriptions. Its financial services have expanded to include residential mortgage loans, credit cards, checking accounts, plus a foray into real estate listings through the "AgLand" real estate program. In addition, TFB has offered its own automobile club, discount travel fares and special-rate long distance telephone service.

J. T. (Red) Woodson, who operated a thousand-acre farm near Gober, growing cotton, wheat and feed grains, as well as running about a hundred head of Hereford cattle, was elected president of Farm Bureau in 1970. His tenure was marked by expansion of group purchasing and commodity activities

Some of Texas Farm Bureau's largest service increases overall, in fact, grew out of the Expanded Program of 1971. Still nebulous when it was approved by voting delegates at TFB's first special called meeting in March of that year, the Expanded Program leaned heavily toward the formation of special-interest "commodity divisions."

Under the leadership of President Woodson and Executive Director Warren Newberry, these developed pretty much as envisioned, coordinated by committees made up of members interested in a specific commodity such as cotton, cattle, sheep and goats, feed grains or rice. The committee approach allowed representatives of those commodities to meet and compare notes on diseases, treatments, government developments, supplies, prices and a host of other factors affecting their interests. As pertinent issues arose, the appropriate committees could explore ways to deal with them and coordinate recommendations for updates to Farm Bureau policy, if need be.

The sheep and goat division, for example, helped coordinate Farm Bureau policy on predator control, diseases such as scrapie and the unsuccessful fight in 1993 to retain federal wool and mohair incentive payments under the National Wool Act. Committee members in the cotton division, likewise, were a driving force in the creation of a statewide boll weevil eradication program, and the cattle committee has dealt with issues ranging from brucellosis and fever ticks to the consequences of massive packer consolidation and the resulting reduction of market outlets.

Another major focus of the 1971 Expanded Program was a commodity marketing proposal that sought in essence to revive an array of marketing operations begun during the 1960s. The 1971 plan was controversial in that it called for putting considerable money into commodity marketing, and that genuinely frightened many members. Other opponents were less frightened than concerned that their dues would be used to promote commodities they didn't even produce. The plan envisioned marketing programs that would be self-sustaining once they became established; the Farm Bureau money was only to be used in the initial stages, to correct a funding deficiency that had been the main cause of the failure of the 1960s programs.

Walter Meyer, Farm Bureau's former director of market development and commodities, fought those early battles, and he remembered them well in later years.

The 1960s programs were conducted under yet another "affiliated" company, the Texas Agricultural Marketing and Development Association. Meyer assumed his Farm Bureau post in 1961 as "pretty much a novice in the field of marketing." But then, so was TFB. "It was all a real new thing in Texas," Meyer recalled, "and from the beginning the program was hamstrung by lack of financing.

"The bottom line was that there just wasn't the money there ... because, let's face it, when you get into marketing there are big bucks involved, and you nearly have to take title to the goods to be sure you've got them."

A case in point was the livestock marketing program. "Even though

Texas Farm Bureau worked hard to establish a livestock marketing program.

we had a contract to sell for them," Meyer explained, stockmen were apt to pull their animals out of the Farm Bureau program and sell them elsewhere if the market improved or if they just got to feeling lonesome for the auction ring."

Much the same thing happened with fruits and vegetables. "We had problems getting everybody organized," Meyer said, "because we had no organization much to do it with. It was very difficult for the Farm Bureau to sustain this program because we were not willing or could not take title to the goods." As a result, "when we wanted to ship strawberries or onions or something, we would work through someone like Griffin and Othal Brand."

Othal Brand, longtime McAllen mayor, was "a real force" in the Rio Grande Valley, and proved to be a valuable ally to Farm Bureau. Meyer recalled a time when onions "were just rotting in the field, and we got a couple of ships in there to take some to Europe in the early spring." That development itself boosted onion prices a dollar "overnight," and suddenly farmers were trying to back out on the export deal to sell to others at the new price. In a sense it was like a snake swallowing its own tail, but Meyer did not have the luxury of enjoying the irony; he had ships to fill.

Othal Brand came to the rescue.

"He said, 'No, we're going to ship this load,' and he personally took it on himself and kept the group together," Meyer said with admiration. "If we had the money," he added, "we would have bought them ourselves" and avoided the problem of reluctant contractors. "But there were several million dollars involved."

Lack of funding to take title to commodities wasn't the only problem Farm Bureau faced, of course. They were up against everything from disinterested middlemen to outright chicanery on the part of competitors or their agents — including foreign governments. But even then, more money might have enabled TFB to build an organization strong enough that it wouldn't have been so easy to stiff-arm.

Meyer told of episodes in which shiploads of fresh grapefruit, already sold, arrived overseas at about the same time as several other loads "from South America or Israel or somewhere." The competing fruit, shipped on speculation, drove the market down enough to leave the Farm Bureau produce overpriced by comparison and the buyer looking for an excuse to turn it down. A phone call would awaken Meyer at three o'clock in the morning, the voice at the other end saying, "Well, your load of grapefruit has gotten here to Amsterdam, and it doesn't look good. There's something wrong with it."

Meyer would ask what was wrong, and "they wouldn't be specific." He would later find out about the competing product, and put two and two together.

Shiploads of fresh grapefruits were being sent to Europe as part of the commodities marketing program.

The Farm Bureau program included some beef shipments to Europe as well as fruit, and those met with resistance, too. The meat was shipped as hanging carcasses in special refrigerated containers, and one shipment actually arrived with the refrigeration units switched off. Other

Joe Baker, from TFB's marketing department, proudly displays Texas produce in a German market.

resistance efforts were less overtly criminal, if no less financially destructive. German inspectors in particular, Meyer recalled, were experts in finding excuses to turn down loads of U.S. beef. When that would happen, Farm Bureau was usually successful in getting it accepted at one of the U.S. military bases in Germany. That was not a sure bet, however.

"We had talked to one of the Army procurement boys," Meyer said, "but they weren't always interested in working with us because they had their own little programs going." Eventually TFB's beef export program attracted such unfavorable reaction among its erstwhile European competitors that the Common Market imposed a 100 percent tariff on it. That was a crippling blow, as intended, and the program folded.

Farm Bureau also tried a slew of domestic marketing programs. One of the things they discovered was that there simply was not enough of an infrastructure to sustain many of them. Meyer described an effort to market cantaloupes from near El Paso to a Safeway warehouse in Dallas. The farmers were eager, "but they just didn't have the packing know-how, and they didn't have hydro-coolers and things to cool them down to where they would keep a long time." The one or two loads the program shipped were already too ripe for the stores when they hit the warehouse. Better preparation and equipment could have avoided that, but it would have cost much more money than either the farmers or Farm Bureau were prepared to spend.

"It always boiled down to the fact that if we wanted to work with farmers on these deals, we'd have to get in and set up a co-op, and there were co-ops there already Meyer said."

Farm Bureau also operated a feeder calf marketing program for a time, gathering calves primarily out of East Texas and selling them to Panhandle wheat and stocker cattle producers. TFB field man Roy L. Custer became involved with that project while serving the Northeast Texas district from 1957 to 1970. From the beginning, Custer recalled, it was a matter of trying to reconcile opposing parties in a dispute while representing both sides.

"The East Texas cattlemen wanted more money for their calves," he explained, "and the West Texas boys wanted to pay less." The catch was that both sides often were members of Farm Bureau, unlike marketing arrangements that involved a producer on the one hand and a packer, processor or retailer on the other.

"The thing about the whole situation is they finally came to realize that each one must give a little to gain a little."

The program had its start in Van Zandt County, Custer remembered, and it was a straightforward gathering arrangement. County leaders would inspect members' calves, and when they received an order for a load, they would contact enough producers to make up the necessary number. "No calves would be brought in to be put on a truck unless there was a particular order for those calves," he said.

The program became big enough and ran long enough for the Van Zandt Farm Bureau and some others to build their own sets of pens, equipped with scales. Then, Custer said, it got a little too big. "The program was a success until the cow jockeys got after it," he explained, "and tried to grow too fast. They took the program away from the county leaders, and after a year or a year and a half, it failed.

"As a field man," he continued, "I thought the program was excellent. My county leaders thought it was excellent, and they were willing to work in it, but we got some people who thought it should go faster. Farm Bureau is noted and has always been noted for moving slowly. This was one of the times we tried to move fast and it didn't work."

As fate would have it, Custer later transferred to the Texas Panhandle just in time to become involved in another feeder cattle program that also failed in 1974. Later Custer helped start a hog marketing program. Like the calf marketing operation, the hog program involved pooling supplies to fill sizeable orders with uniform animals, thereby serving the needs of the packer and earning a premium price in exchange. The hog program lasted several years, until a combination of declining production and producer loyalty, as well as improvements in more traditional markets, led Farm Bureau to phase it out. Its demise was a classic example of TFB faithfully following its own dictum not to serve a need that no longer exists.

Yet another Texas Farm Bureau marketing plan was rather humorously dubbed the "spent hen program." Worn-out or "spent" breeding stock are a natural byproduct of any animal agriculture industry. Cows and bulls that have outlived their usefulness have salvage value in canned or processed meat products, and packer sheep and goats, though too flavorful for most American palates, are well regarded in the Mexican market. "Spent hens," on the other hand, have traditionally been an underutilized asset accruing from the egg-laying industry. Farm Bureau put together a marketing program for its egg-producing members, and countless cans of name-brand chicken noodle and cream of chicken soup contained Farm Bureau chickens that had laid their last egg. Eventually that market, too, developed competition from fulltime people who could devote more time, money and resources than TFB.

At one time, Farm Bureau provided a hay testing service that would evaluate the nutritional content of alfalfa samples. It wasn't a marketing program, per se, but it helped hay producers better market their own product.

Other TFB marketing programs have included turkeys, pecans, carrots and mohair, and the organization has considered additional projects that, for one reason or another, never got off the ground. A broiler program, for example, was deemed infeasible because of the strong hold of certain giant poultry firms over that entire industry.

A 1966 TFB market-builder tour departs for foreign shores.

A truck unloads grain for export.

Roy Custer recalled working with former TFB research director Bill Wedemeyer on a grain marketing proposal for the Panhandle. After two years of careful study, the pair determined that such a program could not work unless farmers could be induced to consign a given share of their crops to the marketing entity for sale on speculation. "Consigning a portion of their grain crop to be marketed without a sold price is something the farmer couldn't quite understand," Custer remembered, "and the grain marketing program never got off the ground."

Texas Farm Bureau's many marketing projects had all been phased out by the 1980s, although no one has ruled out a return to the marketplace, and TAMDA remains as a legal entity should that happen. Passage of the North American Free Trade Agreement in 1993 spurred discussions of possibly targeting trade with Mexico.

In 1991, the Texas Farm Bureau Board of Directors embarked on a four-day, fact-finding tour to the interior of Mexico. The tour focus was to learn more about Mexican agriculture and the potential effects on both Mexico and Texas producers from a free trade agreement.

The board, led by President S.M. True, ran the gamut of Mexican agriculture—from the very small *ejidatario,* where poor farmers try to scratch out a living on small government-owned plots...to small privately owned farms, where diversification of crops and livestock seem to rule...to large operations that could compete with any in the United States.

At stake was NAFTA's purpose of building a 380 million person economy between three nations—Mexico, Canada and the United States.

Jesus Gonzales Gortazar, president of the National Confederation of Private Producers, told the board, "The challenge on both sides of the border will be a cooperative relationship." Mexico and the United States, he said, have complementary economies with Mexico strong in tropical fruits, melons, beef cattle and vegetables. But Mexico, he pointed out, is weak in producing cereal grains, dairy products and oilseeds—a U.S. strong point.

True said the trip did much to dispel misconceptions he held about Mexican agriculture. "It's very impressive," he said. "They can grow crops in Mexico just like we can in the United States. We've seen some very efficient operations and certainly we've seen some very, very small operations that don't compare with what we do in the United States."

He said one sticking point in his mind was that Mexican producers have the same concerns U.S. producers have about free trade as far as markets and overproduction are concerned. He said, however, a free trade agreement would definitely be a net gain, not only for Texas, but for all parties involved.

True pointed out, "I didn't know what a city of twenty million people looked like until we came into Mexico City. There's a lot of mouths to feed there, and it takes a lot of production from all over the Americas to provide that much food and fiber. As we raise the economic level of the Mexican people, and they have the ability to buy more food and more products, I don't think Mexico can supply it. I think we are the closest supplier that they have, and I think we can fill that need for them."

Meanwhile, Farm Bureau's marketing programs by and large did what they were intended to do: they provided outlets when and where existing outlets were inadequate, they spurred competition in many cases where there was little or none, and they generally made money for the Farm Bureau producers they served.

As time went by, Walter Meyer said, "more and more it seemed like our role was to influence the strength of the market, do things that would increase the value of the products, and just help the whole overall marketing program."

If the "old" Farm Bureau's cotton cooperative had managed to maintain that kind of record, there might never have been a "new" Farm Bureau.

Just as the expanded program developed a marketing plan so essential to the future and growth of Texas Farm Bureau, President S.M. True's Blue Ribbon Goals Committee, following the 1985 state convention, became a blueprint to guide TFB for the next decade.

Former state Director James Ray Adams, of San Patricio County, and his sixteen-member committee were given a charge to draft a plan for the future. And with that plan developed, the future was bright.

But the future was also nearer than it had ever been before.

Texas Farm Bureau: The Harvest of Education

Texas Farm Bureau's earliest link with education, the extension service responsibilities on which the Farm Bureau movement was originally based, was severed in the early 1950s by federal legislation that divorced the government-supported Extension Service from private control. By that time, however, TFB had embarked on an educational program of its own that would expand significantly with the passing years.

Texas Farm Bureau's first independent foray into teaching followed the adult education thrust established by Extension programs. As long as there had been field men with TFB, part of their job was to train county leaders in such basic chores as membership record-keeping and accounting. In the summer of 1949, with the number of organized counties growing and the complexities of county-level activities increasing, TFB unveiled its first statewide training workshop for county leaders. Known as the "Farm Bureau Institute," the program was conducted on the campus of Texas A&M. In addition to nuts-and-bolts training, the Farm Bureau Institute provided motivational guidance to help prepare county leaders for the all-important task of recruitment; it prepared them to go back home and "sell" the Texas Farm Bureau.

Former TFB district sales manager Dee Knox was there, and he vividly remembered it many years later.

"We stayed in the old dormitory," Knox said. "It was a battle as to who got the best room." Despite the swamp-like climate of College Station, the A&M dorm at that time had no air-conditioning, not even any fans, and there were no single rooms, so county leaders and field men alike had to "buddy up" with someone they could tolerate, or risk being paired with someone they couldn't.

Knox paired with fellow field man Leon Lane. "You had to get there early to get a good room, so Leon and I rushed down there one afternoon and got a room on the top floor. We thought it would be the coolest room. After we got situated and the sun came out it was hot. That top floor room got the first sun rays and happened to be the hottest room in the building, but we had to stay there because we chose it."

The Institute being aimed at county leaders, TFB's small corps of field men had some time to kill — or so they thought. Knox recalled that he and Lane decided the first afternoon to slip off and play a little golf. "We didn't think we had anything to do around the Institute. When we got back we found out they had called on Marvin Carter to introduce his new field staff, and Marvin must have been pretty embarrassed when he found two out of the three of them were on the golf course."

All was not fun and games, even for the field men. Knox sat in on one speech course intended to teach county leaders how better to present the Farm Bureau story. Walter Hammond had imported the instructor all the way from Missouri, a renowned public speaker who had literally written the book on the subject. As an incentive for better participation, the teacher announced that he would give an award to the best pupil in the class.

"As luck would have it," Knox remembered, "he said I was the outstanding pupil, and he presented me one of the books he had written on public speaking." That impressed Hammond, and a couple of years later the economy-minded Farm Bureau president cajoled Knox into teaching the course himself, writing the costly Missouri professor out of his plans. After all, Knox had the book. He taught the class for years to come.

The Institute was patterned after the American Farm Bureau Institute, which trained field men for the states, just as the Texas version trained on a county level. Longtime West Texas field man Ed Cumbie recalled that he and his colleagues attended the out of state program "as an entire field staff for years and years." AFBF, he said, combined "people ... brought in from outside Farm Bureau to give us motivation" with sessions where fellow field men "exchanged ideas" among themselves. The latter particularly impressed him.

"I think this was experience," he explained, "that you couldn't buy anywhere else, because it was people exchanging ideas and information that we had gained in the field."

Much the same could be said for the state Institute, of course, which ran five years before the statewide concept gave way for a time to area training.That program lasted from 1955 to 1959. TFB then returned to the annual statewide Institute concept through 1965, when the program ended.

Farm Bureau, under president DeVaney, extended its adult education efforts in 1964 when the TFB insurance companies and the safety department teamed up to sponsor training for members of rural fire departments at the annual Texas A&M Fire School. TFB Safety Director Joe Smetana had been assisting the A&M program as an instructor for two years at the time, and the school itself was already thirty-five years old. Like many other Farm Bureau programs, it was ambitious from the start. The state organization and its insurance affiliates would provide most of the funding to sponsor 100-140 volunteer firemen per year, one per county, with the appropriate county organizations matching a modest amount. In addition, the program called for the insurance companies to contribute $75 per incident to volunteer fire departments answering calls on the property of their insured.

It was an example of enlightened self-interest at its best, an organization benefitting the community while looking out for itself. Naturally, went the reasoning, more qualified firefighters and better-supplied and maintained departments would reduce the risks of the insurance companies and lower their claim costs. Farm Bureau recognized that many rural fire departments operated on a shoestring budget and could not afford to send their personnel for proper training. In addition to the property damage the program has helped to reduce, it has undoubtedly saved the lives of at least a few volunteer firemen.

The A&M Fire School and Texas Farm Bureau's participation continue today. The school is conducted at Brayton Field, near Easterwood Airport, part of the A&M University system. Each course lasts a full week, and is usually conducted the last week in July, when students can get a good warm workout even before they approach a fire.

Bridging the gap between adult and youth programs was Texas Farm Bureau's Young Farmers and Ranchers Advisory Committee, formed in 1962. Inspired by C.H. DeVaney, the committee was created in recognition of the need to bring new blood into Farm Bureau leadership, and to make sure the new leaders were prepared to lead once they were given the opportunity. As its name suggests, the committee gave selected young farmers and ranchers an opportunity to "advise" other committees and the state board on issues of the day as seen from the perspective of young adult members.

The youth program was strongly supported by Warren Newberry, who, as a young farmer and rancher himself, knew first hand the importance of young leadership. Newberry had graduated from Lone Wolf High School in Oklahoma and was attending the University of Oklahoma when his father became seriously ill. Newberry left school and returned home to take over the farming and ranching operation.

One afternoon in 1944 he was carrying two buckets of milk back to the house to run through the separator when the president and vice president of the Oklahoma Farm Bureau drove up to the farm house. They told the young man they wanted him to go to Chicago to the American Farm Bureau Federation convention to find out what he could about the national program for young adults.

"We're going to help you with expenses," they said. "The board has set aside fifty dollars to help you."

Even in those days, fifty dollars would not get anyone from Oklahoma to Chicago. Newberry and his mother borrowed money from the bank to buy luggage and a train ticket. He attended the convention and came home with the information and enthusiasm necessary to establish the program in Oklahoma. He became chairman of the Southern

Youth programs are instrumental in helping 4-H and FFA members show their sheep at the State Fair of Texas.

Region of thirteen states. And, after he had completed his term, Newberry was invited by Allen Kline, president of the American Farm Bureau, to direct the Youth Activities for AFBF.

He had always thought he would be a farmer and rancher.

Farm Bureau changed his mind. He spent the rest of his life working for and helping farmers and ranchers.

In return for their contributions and fresh ideas, the committee members receive invaluable insight into just how Farm Bureau leadership functions. If and when they become Farm Bureau leaders, they enter not as green hands but as veterans with experience under their belts. A number of TFB leaders in subsequent years have been "graduates" of the Young Farmer and Rancher Advisory Committee.

Texas Farm Bureau's first foray into youth education was a scholarship program, begun in 1951 and still in place today, that recognizes outstanding leadership in the state's 4-H, FFA and Future Homemakers of America clubs. Early awards were only for 4-H and FFA standouts, but the FHA scholarship was added in 1963. The first awards were for $250 per student, and in 1964 that figure was doubled. It doubled again in 1978 to $1000, and in 1983 increased to $4000, payable at $500 per regular semester over a four-year period.

Farm Bureau is one of several organizations and companies that offer similar scholarships to students in the three clubs. The clubs themselves establish qualifications and standards, subject to TFB's general guidelines (the FFA scholarship, for instance, goes to the state FFA president each year). The Texas Education Agency selects the FHA recipient, and the Texas Agricultural Extension Service chooses two 4-H scholarship winners. Presentations are made at the state FHA and FFA annual meetings and at the annual 4-H Roundup.

The scholarships are intended to increase participation in the three organizations, and, in the process, emphasize leadership in agricultural and rural life.

TFB has not been disappointed in those goals. Typical of Farm Bureau's scholarship recipients are Mark Ellison and Aaron Alejandro, both selected in the FFA category.

In appreciation, Ellison wrote TFB that "the money was important, but the values and traditions of the Farm Bureau that went along with the scholarship were of utmost importance. Everywhere I've gone, whether in this great state of Texas, around the United States or around the world, I have carried a little bit of the Texas Farm Bureau's membership, faith and commitment with me." He termed the scholarship program "a true investment in the future and an opportunity to help rural youth.

"We can have a direct influence in that future," Ellison continued, "by helping them obtain a college education and maintain the strong agricultural, community and family values that are so strong here in Texas on into their professional careers and personal lives."

The product of a Rosebud, Texas area ranching family, Ellison completed his education at Texas A&M and worked his way to the post of senior vice president of a government affairs consulting firm dealing with international trade policy. He served in Washington, D.C. and Casablanca, Morocco, working in more than a dozen African nations before joining the Texas Department of Agriculture, where he became assistant commissioner for marketing and agribusiness development under Agriculture Commissioner Rick Perry.

A resident of Cal Farley's Boys Ranch near Canyon, Aaron Alejandro struggled with the very real problems of adequately funding his own college education, and the Texas Farm Bureau scholarship became an important element in that struggle. "Little did the Farm Bureau know," he wrote later, "that they were investing their money with a young man whose life perspective had been honed with the experience of traveling Texas."

With his education secured by TFB and other grantors, Alejandro continued, "I was able to focus on my studies instead of my finances," and "I began to see the merit of public service and active citizenship in our democracy, never forgetting that my scholarship from the Texas Farm Bureau was for agricultural leadership."

Alejandro worked briefly for the Texas Department of Agriculture, then as a marketing representative for one of the nation's largest agricultural science videotape producers before joining the staff of U.S. Rep. Bill Sarpalius. He became district director for Sarpalius, in charge of four offices in the congressman's district. His civic involvement includes directorships with the Wichita Falls Rotary Club, Boy Scouts and Boys & Girls Club.

Governor John Connally meets with TFB President C.H. DeVaney and a Texas Farm Bureau queen.

"The Texas Farm Bureau has played an integral part in my personal development, both academically and professionally," Alejandro wrote. "I hope I can continue to be the example Farm Bureau promotes in their scholarship investments."

In later years, Texas Farm Bureau added several other scholarships to its youth program. Winners of the annual Talent Find and Miss Texas Farm Bureau contests each receive $2000 scholarships. Runners-up receive $1500 each and thirteen district winners are awarded $600 scholarships.

The Miss Texas Farm Bureau contest had its origin in 1948, when Betty Louise Tally of Robstown was selected Farm Bureau Queen from a field of thirteen contestants representing TFB's directorship districts at the time. Crowned at the state convention by president Walter Hammond, she received a $500 check to defray expenses for her and her mother to attend the American Farm Bureau Federation convention the following month in New Jersey, where she would compete for the title of AFBF queen. Converted to a scholarship program in later years, the contest progresses from county through district levels before professional judges choose the winner at TFB's annual state convention.

Roxie Hillje-Peterson, 1992 Miss Texas Farm Bureau, wrote that the effort she put into several attempts at the crown helped her "learn more about myself and Texas Farm Bureau." She expressed her gratitude to the organization as a whole and to its field staff, with whom she worked to reorganize the contest she had won. The staff, she said, taught her "that a handshake is still more binding than a lawyer's contract will ever be. Farm Bureau sets the type of example that the youth of today look up to and admire."

Marla Rudd, Miss Farm Bureau 1994, said her activities in "this incredible organization" helped her gain "knowledge that is irreplaceable." Farm Bureau, she added, "has been a part of making me who I am today, and I will reap the benefits for years to come. I only hope that many others will allow Farm Bureau to work for them in all the ways it can. May it enrich the lives of people for generations to come."

In 1991 Farm Bureau employees and the family of Dick Mitchell introduced a $750 scholarship to honor the memory of a man who had invested so much of his life to working with those rural Texans who comprised TFB's membership, dedicating most of those years to Farm Bureau youth programs. Mitchell served as a field representative for twenty-eight years until his promotion to field services director in 1989. He had received his B.S. degree in agricultural education from the University of Tennessee, later receiving an M.S. degree in poultry nutrition from Texas A&M University.

The recipient must be a Farm Bureau family member, but unlike other scholarships, the Dick Mitchell Scholarship is also open to the children of Farm Bureau employees and agents. Through the years, Farm Bureau employees have involved themselves in numerous innovative fund raising activities to financially support the scholarship, including the Dick Mitchell Benefit Rodeo and Bar-B-Que Supper at the Heart of Texas Fairgrounds in Waco. The rodeo featured such events as a Kid's Calf Scramble, the Ladies' Boot Race and Calf Branding, done with paint, of course, as well as Team Penning, Team Roping and a Hay Hauling contest.

Perhaps the most coveted Texas Farm Bureau scholarship is that awarded the winner in TFB's Free Enterprise Speech Contest, judged on the basis of five speeches presented on free enterprise. In addition to a $4000 scholarship to the contest winner, Farm Bureau awards $2500 to the runner-up and $1500 each to four finalists, as well as $1000 each to thirteen district winners. The group of six finalists receives a four-day, three-night expense-paid trip to Washington, D.C. to visit elected officials and watch the legislative process in action.

Ben Starr, 1993 speech contest winner, considered himself "extremely fortunate" for the opportunity to compete. In preparation for the contest, he wrote, "I spent many hours looking at our nation's past, learning the principles of the free enterprise system and how it works for the benefit of our people. This wealth of new knowledge has given me an insight into and respect for our wonderful nation. It also helped me to realize that some people do not have this respect, and thereby endanger the system and even our freedoms.

"A wonderful aspect of the Farm Bureau's youth programs," Starr continued, "is that they FORCE you to come into close contact with other people and therefore develop relationships with a wide variety of individuals. This is vital in preparing the way for a future in leadership, which the Farm Bureau stresses so heavily. These friendships serve to enrich your life and make you a better individual.

"Looking back," Starr concluded, "I can think of no other program I have participated in that has made a greater impression on my life than the Free Enterprise Speech Contest. By expanding my friendships with others, enhancing my speech-writing skills, and enriching my knowledge

A young man participates in Texas Farm Bureau's Free Enterprise Speech Contest.

of our heritage, the Texas Farm Bureau has indeed served to broaden and better my life in many ways."

Contest winner Curtis Childers compared Texas Farm Bureau's interest in youth to the story of an elderly man who struggled to cross a huge and almost impassable gully, then turned back "to spend months building a great bridge. When asked why he would bother building the bridge after he had crossed, he replied, 'For one day, a young person will cross here and that person's journey will be much easier because of my contribution. If I cannot leave something to better the next generation, then what good was it for me even to cross?'

"Such is the philosophy of the Texas Farm Bureau and all their youth programs," Childers said. "I am but one of many thousands whose journey has been much easier because of the bridges TFB continues to build."

Despite all of Texas Farm Bureau's efforts to support youth and education, some county leaders felt that more could be done, and in July of 1990, TFB incorporated the Texas Farm Bureau Youth Foundation, a non-profit charitable affiliate, to finance additional scholarships and agricultural research grants. The foundation is supported by private, tax-exempt contributions, and potential recipients must be from families that have been Farm Bureau members for at least two years. Recipients are selected by a scholarship committee comprised of faculty and staff from colleges and universities in Waco.

The Texas Farm Bureau Youth Foundation awarded its first scholarship in 1994 to Paula Evans of Gatesville, who graduated as valedictorian of her high school class. She told Farm Bureau that the scholarship was "definitely a blessing from God," in that she had been concerned about how to finance her college education.

A prerequisite for both the Free Enterprise Speech Contest and the Dick Mitchell scholarship is participation in Texas Farm Bureau's Citizenship Seminar. Since 1963, TFB has hosted four hundred Texas youths annually for a week-long study of the United States Constitution and the free enterprise system. Its stated purpose is to "convey to the participants a better understanding of our American heritage and the capitalistic private enterprise system that has made ours the greatest of all nations."

Students entering their junior and senior years in high school are eligible to be selected by their county Farm Bureaus, and must rank within the upper thirty percent of the class in scholastic standing.

The first Citizenship Seminar was conducted on the campus of Sam Houston State University in Huntsville, but organizers found out quickly that the East Texas site was inconvenient for would-be participants from the western part of the state and the Panhandle. With the Farm Bureau headquartered in Waco, Baylor University was a natural location for the seminar, and it spent the next two decades there. Eventually, however, the centrally located Baylor campus attracted too many summer conferences, and the college was forced to limit itself to church-affiliated activities.

Ed Cumbie was involved in the seminar from its inception, and his friendship with administration officials at Angelo State University in San Angelo provided a solution to Farm Bureau's dilemma. Cumbie recalled that the late Dr. Lloyd Vincent, longtime ASU president, "became very enthusiastic about the program" when he learned what it was all about. "They agreed that they would like to sponsor it if we wanted to move there, so we did."

Cumbie, who participated in twenty-six seminars before his

Participants come from all over the state to attend the TFB Citizenship Seminar.

retirement as a field man, termed the program "something that we can certainly be proud of, something that Farm Bureau could long be famous for and certainly could be commended for sponsoring." Participants, he added, "are exposed to speakers and situations there that you couldn't buy in college and you could get nowhere else."

"Graduates" of the seminar have gone on to make names for themselves in a number of fields, including government. Alumni include former State Senator Cyndi Krier of San Antonio, who said the seminar "left an indelible mark early in my life that provided an impetus for future political endeavors." She encouraged others "to take advantage of the opportunity being offered, and more importantly, make the most of it. You'll never know where it might lead without first venturing to discover the possibilities."

Other participants who went on to make names for themselves include U.S. Rep. Jack Fields of Houston and Fred D. McClure, former Assistant to the President for Legislative Affairs under the Reagan and Bush administrations.

And then there was Vernie Glasson, who attended the first Citizenship Seminar in 1963 as a sixteen year-old senior-to-be and went on to become Texas Farm Bureau executive director nearly three decades later. In between, Glasson put his seminar experience to good use with a political science degree and a stint during the 1970s as first assistant and then director of the American Farm Bureau Federation's legislative and regulatory affairs in Washington.

Glasson only "happened" to attend the seminar in 1963, he told an interviewer later, "because the fellow who was first selected to go couldn't go. I was really second choice, but when I went I really fell in love with the Farm Bureau and with a lot of the people, and certainly the philosophy and the thought behind Farm Bureau. I decided then that I wanted to go to work for the Farm Bureau and that I wanted, as a career, to be a Farm Bureau field representative.

"Sometimes," Glasson remarked, "you just happen to be in the right place at the right time."

Up to the 1980s all Texas Farm Bureau educational programs dealt with young people from Farm Bureau families, but in 1985, with S.M. True heading the organization, TFB made a major departure from that practice, adopting the "Agriculture in the Classroom" program initiated in 1981 by USDA and placing it under the direction of Glen Jones. Jones had replaced Bill Wedemeyer as head of Farm Bureau's Information, Research and Education Department. He had continued writing and producing the popular "*Facts for You*," but putting the column twice a month in *Texas Agriculture* instead of mailing it out weekly. Now his attention turned to educating Texas youth as well. For years, agriculture leaders and the producers at home alike had fretted over the realization that children — and, increasingly, adults — no longer knew where food comes from.

The once-humorous observation that city kids thought milk came

Ag in the Classroom gives students an opportunity to better understand the various facets of food and fiber production in the State of Texas.

from a carton and hamburger meat from McDonalds had become less and less humorous as the years passed, because those same ignorant youngsters were taking up positions of power and influence as adults. More and more, government policy on matters crucial to agriculture was being made by people who couldn't tell milo from mashed potatoes, and the results were beginning to hurt. From hysterical overreaction to the latest scare story about pesticides to interference in research and production, uninformed activists, government bureaucrats and the media alike were strangling production agriculture.

"Agriculture in the Classroom" was an effort to address that ignorance by making educational material available on agricultural production, processing, transportation, manufacturing and research. The program supplies teachers with factual information in the form of resource guides and videotapes as well as speakers and field trip opportunities. County Farm Bureaus are responsible for disseminating AITC materials to schools in their area, and as of 1994, fully sixty percent of Texas schools had the material available, keyed to the fourth grade level, where information about Texas is taught. In addition, guides for first through third grades are also available, and new materials are being developed as necessary.

Jones developed the resource guide, along with other supplementary materials, giving teachers an opportunity to guide students as they explored various facets of food and fiber production in Texas. In addition, County Farm Bureau began making it possible for local resource people to explain area farming and ranching; provide animals, crops or other farming and ranching demonstrations; hosting farm tours and arranging agri-related tours where they were feasible; visiting area museums which preserve and demonstrate the state's agricultural heritage; and corresponding with students about what life on a modern day Texas farm and ranch is really like. The resource people include county agents, the soil conservation service, high school vocational agriculture teachers, representatives from experiment stations, feed lot operators, tractor and implement dealers, farmers and ranchers.

Jones pointed out, "The shift from a rural to an urban society has left the general population with little knowledge of the major industry that provides food and fiber to the nation. Farm Bureau's aim with Agriculture in the Classroom is to acquaint urban students, their parents and teachers with the major role that agriculture plays in their lives. We want to demonstrate to them how vitally important Texas agriculture is, not only to them personally, but also to the United States and the world.

"Through Agriculture in the Classroom, many more young people can be exposed to the strong moral values and work ethic that farm and ranch families are known for, even make them aware of career choices available in the field of agriculture."

Farm Bureau works with such Agricultural organizations as

The Ag in the Classroom program provides educational material on agricultural production, processing, transportation, manufacturing and research.

Agri-Food Masters, Texas Agri-Women, Texas Sheep and Goat Raisers Association and Texas Cattle Women to promote and distribute Agriculture in the Classroom materials to schools throughout the state.

The Ag in the Classroom program was given a boost in 1991 with a $45,000 grant from Houston Livestock Show and Rodeo. With the money, Farm Bureau was able to provide resource material to fourth grade teachers in seven county areas surrounding Houston. The San Antonio Livestock Exposition and the State Fair of Texas have also been extremely supportive of the Ag in the Classroom program.

A complementary program initiated in the summer of 1994 is the week-long Summer Agriculture Institute, a seminar for Texas educators that begins with a basic introduction to Texas agriculture, proceeds through hands-on training, a tour of agriculture processing facilities and a panel discussion on agricultural issues. The seminar is conducted on the Tarleton State University campus at Stephenville and earns participants three hours of graduate credit through Tarleton State.

Texas Farm Bureau's education programs continue to expand, and include farm tours, agricultural field days, and teacher in-service training. Clearly, the breaking away of the Extension Service program four decades earlier did not end Farm Bureau's involvement with education.

TEXAS FARM BUREAU: THE REWARDS OF RESEARCH

The Fight to eradicate the screwworm became one of the most important battles in the history of U.S. livestock production. Although Texas Farm Bureau was not involved directly in the raw research, the organization played a key role in assuring that the results of that research were put to use. The net outcome has been savings of literally billions of dollars by stockmen and consumers alike, in the United States, Mexico and an increasing share of Central America. In addition, there is no way to measure the benefits to wildlife, nor to estimate how many human lives have been saved. Farm Bureau's part in the screwworm eradication effort is justifiably one of the organization's proudest accomplishments.

The screwworm, *cochliomyia hominivorax*, has had a half-dozen or more scientific appellations through the years, most of them translating in some way or other into "man-eater." Barely distinguishable in appearance from the common blowfly, the screwworm is eminently distinguishable in habit; it is the only fly maggot in the world known to feed strictly on living flesh.

The fly was seldom known to winter above South Texas, and severe cold seasons killed them back well into Mexico. With spring, however, they would spread north again, each fertile female producing hundreds of thousands of offspring during her brief life, and those offspring — and theirs — procreating geometrically to infest millions of acres of land.

They were a threat to every warmblooded creature on every infested acre, humans by no means excepted. Charlie Scruggs, in his definitive treatment of the screwworm, *The Peaceful Atom and The Deadly Fly*, recounts many gruesome stories of people infested by screwworms, usually through eggs laid in the mucous lining of the nose. In such cases, the maggots would quickly spread to the sinus cavities, often the lungs, and eventually the brain. Some victims survived, horribly disfigured, but most who contracted screwworms before the days of disinfectants, antibiotics and modern surgical techniques succumbed after the most grievous suffering.

But by far the greatest losses to the screwworm have come in wildlife and livestock. For as long as men have raised cattle, sheep, horses and goats in the New World, they have lost a large percentage of them to screwworms. Castration, dehorning, branding and earmarking all produce wounds, as do shearing and tail docking in sheep, and every wound is a potential screwworm infestation site.

For that reason, stockmen long ago learned to perform necessary surgery after the first good freeze in the fall and before warm weather returned in the spring. But that wasn't enough; the very act of birth carries with it the seeds of death in the form of a raw navel, so many ranchers and farmers adopted calving, lambing, kidding, farrowing and foaling practices at odds with nature.

Despite producers' best efforts and sacrifices, animals still suffered wounds. Bulls and bucks butted heads, animals of all types scraped themselves on brush or wire, and horses seem to be born with the ability to injure themselves even in their sleep. During fly season, each wound was potentially mortal, so stockmen spent an inordinate amount of their time "riding for worms." The unrelenting practice produced good ropers and well-schooled horses, but at a frightening cost. In modern terms, the combination of lost livestock, lost production and the expense of treatment was costing the livestock industry hundreds of millions of dollars a year.

It was enough of a problem to attract federal interest. By 1930, USDA had established a research station in Dallas and satellite outposts at

The dreaded screwworm posed one of the most dangerous threats for horses and cattle all over Texas.

A calf is checked for any sign of a dreaded screwworm infection.

Menard and Uvalde. Their main purpose was to develop improved treatments for screwworm-infested wounds, and that required the trapping and production of flies. One of the Menard researchers, Emory C. Cushing, made the first significant advance in screwworm control when he discovered through microscopic examination that the larvae which fed on live flesh matured into a different fly than that produced by common blowfly maggots. For the first time, the scientists truly knew their enemy.

In 1933, with the Dust Bowl and the Great Depression both pounding the Southwest, the federal government instituted a program to move drouth-stricken cattle east. Somewhere in the confusion, screwworms went along.

USDA responded with yet another screwworm lab, this one at Valdosta, Georgia. One of the Valdosta researchers was transplanted Texan Edward F. Knipling, who grew up with screwworms and had earlier served at the Menard substation. Knipling studied the parasite's life in minute detail, searching for some as-yet unknown chink in its armor.

The next piece in the puzzle came from the Dallas station, where another young entomologist, R.C. Bushland, had begun work in 1935. As low man on the totem pole, it was Bushland's job to tend the "stinkhouse," where fly larvae were raised in the wounds of sacrifice rabbits. It was a disgusting job and a cruel one, though at that point the regrettable cruelty was unavoidable.

Bushland so hated the stinkhouse that he was driven to find a substitute, and he did. After a long spell of trial and error, he discovered a mixture of hamburger meat, blood, water and formaldehyde in which the larvae would thrive. It eliminated the need to sacrifice rabbits and other animals, and made mass production possible in an efficient manner.

Bushland and Knipling were both transferred to the Menard lab in 1937, where their paths crossed for the first time. Their mutual interest in screwworms meshed, and in Bushland's teeming masses of lab-produced flies, Knipling found the subjects for thorough study. Careful observation led him to the conclusion that female screwworm flies mate only once in their lives. What if, Knipling wondered, sparse native screwworm populations could be overwhelmed with lab-raised sterile males? Could the fruitless matings depress overall reproduction enough to eliminate the screwworm as a serious pest? Or even eliminate it altogether?

It was an intriguing thought, but there was simply no practical way to sterilize male screwworm flies in any significant number.

Meanwhile, the attention of much of the scientific world turned to the dawning nuclear age, given dramatic birth at Hiroshima and Nagasaki. Warning of its potential dangers, former University of Texas geneticist H.J. Muller predicted in 1950 that fallout from nuclear warfare could produce a world of mutants — or lead to widespread sterility. He based his claims on experiments he had conducted in 1927 on fruit flies dosed with X-rays, and for which he had received the Nobel Prize.

When Knipling heard of Muller's long-forgotten study, he contacted the scientist, outlining his sterile screwworm theory. Would it work, he wondered? Muller termed the theory "sound," assuming the radiation did not also damage the sterile flies and leave them uncompetitive in the wild.

With his own mass fly-rearing techniques and Muller's detailed notes, still on file at the University of Texas, Bushland had everything he needed to test the sterile fly concept — everything but money. The hospital in Kerrville wanted $200 per test for the use of its X-ray machine, so taking full advantage of an army friendship, Bushland prevailed on a colonel at Brooks Army Hospital in San Antonio. The tests worked, and Bushland managed to scrounge up an old cobalt-60 gamma radiation unit from the Oak Ridge atomic laboratory in Tennessee. He learned to operate it himself, still pinching pennies, and with a small grant confirmed in limited cage tests that the introduction of sterile males into a normal screwworm population could lead to eradication in a few short generations.

With the derisive laughter of old critics still echoing in their memories, Bushland and Knipling chose to put the theory to a test in the real world. Bushland chose the island of Sanibel off Florida's west coast. A few months of sterile fly releases there reduced but did not eliminate fertile egg masses in sacrifice goats. Deducing that fertile flies from the mainland were reinfesting the island, Bushland and Knipling sought another test site out of easy reach of land. Luck intervened in the form of a Dutch agricultural officer on the island of Curacao off South America. Screwworms were decimating islanders' goats, he wrote Knipling.

Cattle were particularly at a risk, as the fight to eradicate the screwworm became one of the most crucial battles in the history of U.S. livestock production.

In the summer of 1954, sterile flies from a temporary plant at Orlando began arriving over Curacao, dispersed by air from paper bags. Once the researchers found the proper distribution density, it took only 14 weeks of sterile fly drops to eradicate screwworms from the island — permanently.

Florida was next. When stockmen in the peninsular state heard of the Curacao experiment, they literally demanded that a similar trial be conducted in Florida; they had not yet come to accept the ravages of screwworms as simply part of nature, and were determined to rid themselves of the scourge one way or another.

Things began to jell in the fall of 1956, when unusually warm weather made screwworms worse than usual for Floridians. Livestock had to be flushed out of the jungle-like cover and treated three or four times, and countless thousands died anyway.

When the cattlemen's group met in January of 1957, directors easily passed a resolution requesting that Congress fund a screwworm eradication program for Florida and neighboring states, and Florida's own legislature took up the cause as well. When stockmen offered to chip in through a per-head checkoff on cattle and hogs, state lawmakers appropriated three million dollars for the project with the provision that it be matched by federal funds.

With a boost from Mother Nature, the frigid temperatures froze overwintering screwworms back to the southern tip of the state, creating an ideal situation in which to launch an eradication drive. The program would not have to blanket the entire state; a relatively narrow buffer zone could stop reinfestation, targeted fly drops could clean up scattered outbreaks north of that line, and blanket dispersal would be needed only south of the buffer.

As of late February 1959, screwworms were officially history in the Southeast. They remained, however, very much a thing of the present in the Southwest, and the effort by that region's stockmen to repeat the Florida success story brought the Texas Farm Bureau into the picture.

Agriculture and sportsmen's groups in Texas and surrounding states had watched the Florida experiment with growing excitement. As crazy as it had sounded at first, the notion of wiping out screwworms with radioactive sterilization had proven itself. The only real question was whether it could be duplicated on a much larger scale in Texas, and whether reinfestation from overwintering areas in Mexico could be controlled in a cost-effective manner; Florida, after all, was surrounded by the sea and faced no reinfestation once it had been cleaned up, but Texas had Mexico and the rest of Central and South America as a breeding ground for reinfestation.

As they had in Florida after Curacao, USDA officials were already looking Texas over as the Florida program showed signs of success. And, as with Florida, their Texas plans were deliberately low-key. To help lay the groundwork for an eradication effort in Texas and the Southwest, officials of USDA's Animal Disease Eradication Division began collecting and reporting the number of screwworm cases in the region.

Meanwhile, Senate Majority Leader Lyndon Johnson weighed in with his support of the plan. Having fought screwworms as a youngster in the Texas Hill Country, Johnson understood the problem on the most personal level. His clout earned serious attention for screwworm eradication when U.S. President Eisenhower and Mexican President Adolfo Lopez Mateo met in February 1959. The meeting led to a joint U.S.-Mexico survey and eventually to a joint eradication program.

For the time being, however, an eradication program in the Southwest was officially off: Agriculture Secretary Ezra Taft Benson opined that such an effort "does not appear feasible" without "further research and field trial," then dismissed such a trial as "quite costly." USDA brass ordered subordinates not to even discuss screwworm eradication in the Southwest, a situation that prevailed through 1960 and into 1961. As late as the summer of 1961, Bushland's superiors called him on the carpet in Washington for meeting with Texas producers who were having one of the worst screwworm years in recent memory.

It was June 1961 when the logjam first began to crumble. Defying his superiors, USDA ADE official Dr. Robert J. Anderson got word to stockmen that Congress might indeed accept a screwworm eradication program — but only if producers themselves helped fund it. That news spurred action by the Texas Animal Health Council, an informal umbrella

group representing nearly three dozen agriculture organizations including Farm Bureau, the Texas and Southwestern Cattle Raisers Association and the Texas Sheep and Goat Raisers Association.

To fight the screwworm, producers were asked to contribute fifty cents per animal unit, defined as goat, sheep, cow, bull, horse or swine.

At a July 21 meeting in San Angelo, Texas Farm Bureau vice president C.H. DeVaney committed his organization, as did representatives of the others, to formation of the Southwest Animal Health Research Foundation, a non-profit organization that would collect and disburse producer contributions to the eradication effort. Participating groups each pledged $500 as seed money for postage, phone calls and printing.

The program was going to take more than seed money, however; any eradication effort big enough to work in Texas, the Southwest and Northern Mexico would require a new fly plant close to the action and with a fly production capacity that would dwarf anything done in Florida.

Based on their Florida experience, Bushland and other researchers concluded that a buffer zone one hundred miles deep would be needed to intercept and neutralize migrating female flies. To eliminate screwworms from Texas and then blanket an area that wide and as long as the Texas border, more than a thousand miles, experts estimated, would cost $12 million a year over three years. Some estimates ran to $20 million.

SWAHRF's first board included Farm Bureau's DeVaney, who proposed that stockmen set their voluntary collection goal at $3 million. To reach that goal, the group would ask producers to contribute 50 cents per animal unit, defined as one cow, bull or horse or five sheep, goats or swine. Dr. Bushland was given the honor of making the first official contribution, the pledge of royalties from a book to which he had contributed.

Texas A&M volunteered the use of its county Extension program committees to get the word out to producers, but legal restrictions on fund-raising activities by government employees limited what county agents could do. Farm Bureau was uniquely qualified to take up the slack; with its grassroots, county-level organizations, it already had a structure in place. In addition to DeVaney, TFB contributed two additional trustees, Paducah cattleman T.J. Richards and Marlin cattleman and veterinarian Gibney "Doc" Kendrick. Both were Farm Bureau directors.

By the fund drive's official kickoff date of Nov. 1, 1961, seventy percent of Crockett County stockmen had already chipped in, collecting more than $27,300. Within another month, 157 of Texas' 254 counties had active producer committees, and sixty days after the drive began, SWAHRF had $1.5 million in various Texas banks.

But the drive wasn't confined to Texas alone. Producers in Louisiana to the east, Oklahoma to the north and New Mexico and Arizona to the west were also lobbied for support. SWAHRF trustees visited tribal leaders on New Mexico and Arizona Indian reservations, and they called on influential stockmen in Mexico.

TFB's Kendrick later recalled "some wild times meeting at some of these ranchers' haciendas and being wined and dined by the Mexican people, because they wanted the screwworm eradicated from Mexico, too. We were trying to cultivate the Mexican people and they were trying to cultivate us, all at the same time."

With a million and a half dollars in the bank, SWAHRF leaders descended on Washington, seeking the federal funding which they believed they had been implicitly promised. It was a rocky trip, and appeared to have been a failure until virtually the last minute when the Texas stockmen wrangled a brief audience with then Vice President Lyndon Johnson. With his help, they reversed the Kennedy administration's previous opposition to a Southwest screwworm eradication program. In early February, Kennedy called on Congress to appropriate $3 million for the effort.

There was a catch — in their determination to sway Washington, SWAHRF officials had not only vowed to double their own current funding level, they had also committed the Texas Legislature to chip in $3 million. The federal money would be contingent upon those matching funds, and then only if Congress could be persuaded to go along.

With that shaky promise to back them up, SWAHRF trustees committed themselves to the construction of a new South Texas fly plant capable of turning out a hundred million sterile flies per week. Original estimates placed the cost at $2.25 million and the construction timeframe at one year. With building shortcuts and the donation of surplus equipment from the mothballed Florida program, however, both days and dollars began peeling away. The final cost estimate was $650,000 and the timeframe only six months. The contractor, Brown & Root Construction, improved even on that. Headed by concerned cattleman George Brown, the construction company came in under budget and two weeks ahead of its own ambitious July 1 dedication date.

That was fortunate in two ways. First, as in Florida, the Texas eradication program received an early boost from Mother Nature in the form of a deep freeze that killed most flies out of the Texas overwintering zone and gave screwwormers a head start. To take advantage of the freeze, however, the program had to be started immediately, using flies from the small Kerrville plant. As the effort progressed, that plant's capacity was strained to keep up, and completion of the new plant at Mission in deep South Texas became essential to avoid loss of momentum.

The second factor was Washington politics, which stalled delivery of federal funds. As a result, the Texas fly plant was actually built and for some time operated on producer funds, and there was none of that to spare.

Eventually both the state and federal governments came through with the expected funding, but in the meantime SWAHRF leaders were forced to go back again and again to their producers for additional money. Even with success in their grasp, government officials repeatedly appeared determined to fumble it away and waste all the time, effort and money contributed by the stockmen of the Southwest. Each time the call went out for stopgap contributions, Farm Bureau and the other organizations rolled up their sleeves, mobilized their local groups, and astounded even themselves with their success and the willingness of livestock producers to dig deep and pay for a program they believed in. When all was said and done, SWAHRF had collected roughly $4 million, a third more than its bold pledge.

By the end of 1963, the program Ezra Taft Benson had dismissed as "not feasible" only four short years before was credited with reducing screwworm cases in Texas by ninety percent, and in early 1964, SWAHRF unofficially pronounced screwworms "eradicated" from the United States. That wasn't the last Texas producers would see of screwworms, not by a long shot, but from then on, all infestations were considered invaders from central and southern Mexico, below the sterile fly buffer zone which stretched across Mexico's northern states.

The buffer did double duty. By freeing producers in northern Mexico from screwworms, it gave that country a taste of what eradication could mean, and it was a taste that, once acquired, the Mexicans didn't want to lose. By 1970, Mexico and the United States were embarked on a joint program to eliminate screwworms from the remainder of Mexico.

Less dramatic, certainly, than the screwworm battle but no less heated at times and about as long-lived has been Texas' war on *bovine brucellosis,* also known as Bangs disease, and in humans as undulant fever.

Cattle had to be rounded up and tested for incidence of the dreaded bovine brucellosis.

Under federal mandates, brucellosis "eradication" amounted to testing cattle and slaughtering anything that tested positive. Producers, particularly in South and Southeast Texas, soon found themselves in what seemed to be a perpetual nightmare of testing and slaughtering; entire herds could be depopulated and replaced, and before long the new cattle would be turning in positive test readings. For those producers, the "cure" was far worse than the disease had ever been.

Texas Farm Bureau was one of the first organizations to embrace the eradication concept, but that was not the group's original policy.

TFB's Doc Kendrick recalled a state convention during Harrell West's presidency when the first seeds of change were sown. He remembered because he sowed them himself. Anti-eradication program feelings ran so high at the time, he said, that it was a "mob-type situation .. I mean, if there was any legislation about coming and testing our cattle, well rule that out because the tests weren't any good and brucellosis was not that big a problem. You just touched a sore spot if you mentioned brucellosis."

A resolution on the issue came up for adoption, and it "sounded to the membership like just exactly what they wanted because it was going to keep us miles away from anything to do with brucellosis." Yet, Kendrick remembered, "it was head-in-the-sand type thinking," worded in such a way that he felt it would prove embarrassing to Farm Bureau once people began to understand its meaning.

At that time, any member was allowed three minutes to speak for or against a proposed resolution, and Kendrick used his three minutes. Afterward, "I saw it was still going to pass and I hadn't defeated it at all, hadn't even made a dent."

In desperation he approached parliamentarian Al Grauer, seeking another chance to speak against the resolution. The three-minute rule was strictly enforced, and Kendrick had used his allotted time, but he convinced Grauer to let him take the mike again, the second time not as a Farm Bureau member but as a veterinarian and thus an "expert" witness. The second try was the charm, and he managed to persuade delegates to amend the resolution enough that it was no longer an "embarrassment."

"From that day on I think the progress and support for the eradication of brucellosis continued to build, because people began to see that it was costing them worlds of money," Kendrick said.

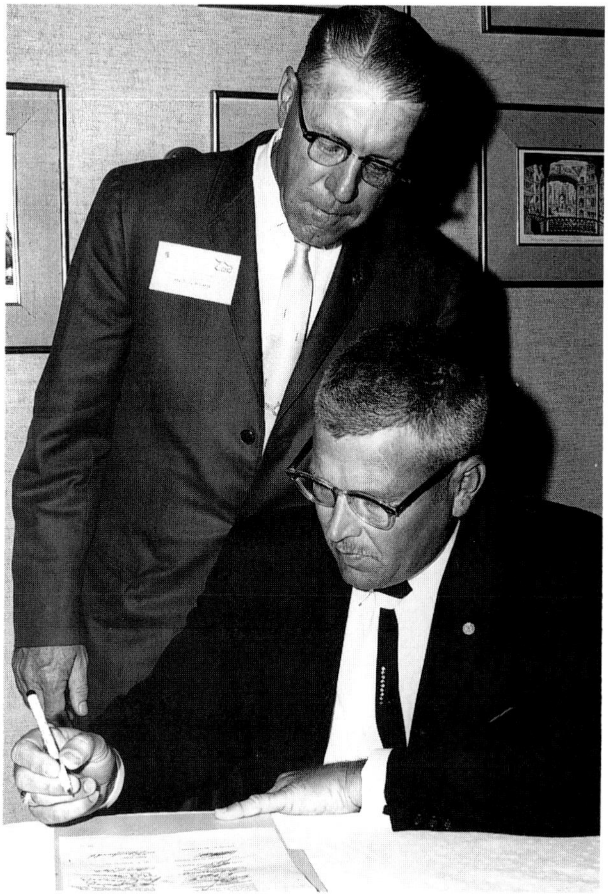

TFB's Doc Kendrick, sitting, with TFB board member T. J. Richards, was instrumental in persuading Texas Farm Bureau delegates to pass a resolution favoring eradication of brucellosis.

With Farm Bureau's backing, the program itself began to change for the better. One of the most promising tools for true eradication had long been vaccination, but the federal government's hard-nosed insistence on test-and-slaughter soured many producers on vaccination because the immunity imparted by the vaccine tended to cause cows to test positive for the disease. It took better tests and official adoption of a reduced-dosage vaccine to induce much support for vaccination, but once that was achieved, eradication moved quickly.

It's perhaps fitting that one of Texas Farm Bureau's latest research-related victories has been over one of the first pests addressed by the Extension-Farm Bureau movement. It was fear of the boll weevil, after all, that led Northeast Texas businessmen to fund the first "method demonstration" projects in Texas. In the early 1990s, Texas Farm Bureau, led by presidents True and Stallman, was instrumental in implementing a coordinated eradication program that may eliminate the boll weevil threat once and for all.

The earliest demonstration projects involved cultural practices to minimize boll weevil damage, but in later years those methods were overshadowed by the emerging technology of pesticides. Instead of crossing their fingers and hoping to either outwit or dodge the cotton pests, farmers found they could kill them wholesale. That early pesticides lost their potency fairly quickly as surviving boll weevils developed resistance was of no serious concern as long as new products continued to surface.

Researchers and producers alike began looking once again at cultural practices, trying to find ways in which crop manipulations could stretch the effectiveness of remaining pesticides. One answer for cotton producers was the concept of uniform planting and destruction dates. By destroying crop residues soon enough after harvest and delaying planting long enough in the spring, farmers could deny boll weevils the habitat they required to overwinter. Carried far enough, the practice could do to boll weevils what sterile male flies did to the screwworm.

As distasteful as the prospect seemed to farmers who valued their freedom and the right to do with their land as they saw fit, most producers gradually came to accept the fact that the promise inherent in uniform planting could never be realized without some mechanism for mandatory compliance. That sobering reality found its way into Texas Farm Bureau policy, and TFB was instrumental in crafting and backing a statewide boll weevil eradication program that was introduced to the Texas Legislature in 1993. It was Texas Farm Bureau's second battle on the issue. The Legislature had passed a bill supporting the program in 1991, but Governor Ann Richards vetoed it.

The program immediately drew fire from environmental activists, allegedly because of concerns that its initial efforts would require stepped-up pesticide use to set the stage for the altered cultural practices to come. Farm Bureau riddled that argument by pointing out that the program would lead to dramatic reductions in pesticide use in the long run, compensating many times over for any increased use early on. Critics continued to complain, however, replacing one complaint with another as necessary. Their hole card was Gov. Ann Richards, who could be counted on to side with them in almost every instance.

By the time the plan came to a vote, however, Richards was reduced to arguing that she could not sign legislation that interfered with the rights of landowners to farm as they pleased. Given her positions on other property rights-related issues, it was a ludicrous stand for Richards to take, especially since the plan enjoyed widespread backing by the very farmers whose interests she claimed to be representing. Richards' opposition finally crumbled, and the Texas boll weevil eradication program became law.

Texas cotton producers had been battling the boll weevil for more than a century, but now they had the tools to eliminate or at least control the pest. The Texas Department of Agriculture would oversee the eradication program, and TDA officials were calculating that, by using certain controls, farmers would need seventy percent fewer pesticides. It was, frankly, agriculture's way of being progressive. As one farmer told Farm Bureau, "What we've got out here now is good bugs eating the bad bugs."

Dan Pustejovsky, a former TFB board member, said, "This was the culmination of more than twenty year's work of getting enabling legislation for us to start the boll weevil eradication. We're still about twenty years behind, but maybe in five or ten years from now we can start reaping some

of the rewards of the program."

If history is any guide — and it almost always is — Texas Farm Bureau will be called on again to defend the boll weevil plan before it accomplishes its goals, and it will be embroiled in other as yet unforeseen struggles down the road.

Texas Farm Bureau worked to pass a Texas boll weevil eradication program in order to protect the cotton farmer's crops.

Texas Farm Bureau: Working for Tomorrow

Texas Farm Bureau has always had its eye on the future. When some were seeking better lanterns, Farm Bureau sought rural electrification. While others dreamed up new words to curse the mud, Farm Bureau haunted the halls of the Capitol in a crusade for better roads. As manufacturers advertised better screwworm smears, Farm Bureau was on the front line in the battle to eradicate screwworms once and for all. From the days when method demonstrations sought a way to live with boll weevils rather than abandon the land, to the recent cooperative effort to eliminate boll weevils entirely, the Texas Farm Bureau by whatever name has represented agricultural producers who have not been content to shake their fists at longstanding problems but who instead have rolled up their sleeves and worked for something better.

As the Texas Farm Bureau passed its sixtieth year, there was quite a future for it to contemplate, a future both of challenge and of promise.

Much of the promise — for Farm Bureau and for the rest of the nation, if not the entire world — lay in unmistakable signs that the free enterprise, democratic philosophy which has driven TFB since the days of Walter Hammond has finally taken root throughout the globe. The fall of the Berlin Wall and the disintegration of the Soviet Union in the early 1990s was the most obvious signal of socialism's collapse, and both Red China and Cuba appeared poised to follow.

At home, by the Eighties and Nineties, the public appeared to have lost patience with the social pathologies which had turned much of American society into a wasteland, one that men like Walter Hammond would not have recognized. America's people were increasingly unwilling to accept the same old empty promises backed up by the same old excuses for failure. As the property rights revolt in Texas amply demonstrated, Americans from all walks of life were also beginning to recognize that denying a farmer or rancher the right to use his land as he sees fit differs only in degree from denying a homeowner or apartment dweller the right to lock his doors.

The challenge facing Farm Bureau in the future is that many of the positive developments of recent years have yet to come to full fruition. What appears to be a rededication to old values and a rebirth of common sense is not nearly complete. Those who have benefitted from the disintegration of American society still hold considerable sway, and pseudo-scientific nonsense still hampers not only the full development of the nation's resources, but in many instances their protection as well.

Biotechnology is a case in point. Texas Farm Bureau policy supports biotechnology and calls for increased research as long as it doesn't interfere with research on traditional agricultural production. The field holds tremendous promise in a world where finite resources must support a still rapidly growing population. With that promise come real questions of economics, proprietorship and even ethics, but by far the greatest controversy has involved fearmongering and essentially bogus questions of "safety."

Texas Farm Bureau had long worked closely with Texas A&M University, and when the focus turned to the study of biotechnology, Dr. Charles J. Arntzen was brought into Texas A&M in a move aimed toward research into biotechnology.

S. M. True recalled, "When Dr. Arntzen arrived, he indicated to most of us that his modern research pertaining to biotechnology would not, in any way, jeopardize what we called traditional research."

But times were changing. And True and Texas Farm Bureau could see, more and more, that traditional technology on various commodities and products was being ignored at Texas A&M, while the exciting new field of biotechnology was being stressed. And they were troubled. Funds were being transferred. People were being transferred. Dr. Arntzen combined extension, research, and veterinary science agencies into one, eliminating each of their administrators.

As far as Texas Farm Bureau was concerned, that was a grave mistake. Rural Texas still needed the research coming from each of the agencies, but traditional research, which farmers and ranchers in Texas needed most, was beginning to suffer.

As a result, True said, "We had to make a change in the administration, in the direction agriculture in Texas was going at Texas A&M. We had to change the people who were involved in heading up the research, as well as the agricultural part of Texas A&M."

In February of 1992, Dr. Arntzen, aware of the dissatisfaction rising up across Texas, resigned as deputy chancellor and dean of the College of Agriculture. He was replaced by Dr. Edward A. Hiler, who was serving as head of the Engineering Department at Texas A&M.

True remembered, "Dr. Hiler's approach to research had a very tranquilizing effect on the concern that Texas farmers and ranchers felt. Under his leadership, we were able to return to a more traditional form of research, doing business the way it should have been done."

The bovine growth hormone BST, for example, promotes an increase in milk production that could help hard-pressed dairymen caught in a squeeze between rising production costs and flat prices for their product. In many instances, increased efficiency with existing inputs is the only avenue left for improving the dairyman's bottom line, and BST has proven itself to be a "silver bullet" in the quest for ever-more efficient production.

One legitimate question is whether or not such improved efficiency can be justified in an industry that is already its own worst enemy, an industry so remarkably efficient that output continues to rise even as cow numbers decline. The dairy industry achieved that level of efficiency through biotechnology of a slightly earlier era, using artificial insemination, embryo transfers and scrupulous progeny records to produce a cow herd and bull pool literally "engineered" to perform to the highest and most predictable specifications. It can fairly be asked if another major improvement will only worsen the existing overproduction and further depress prices, giving dairymen little or no net gain and perhaps even a net loss.

Another legitimate question is whether dairymen can afford to become dependent upon products such as BST, and thus captive to the potential demands of their manufacturers. Opponents of this and other similar new technologies have argued that allowing their use by anyone makes that use virtually mandatory for all, because failure to adopt the technology imposes too great an economic handicap. They are prone to insist that new technologies must be banned to save producers from themselves.

Texas Farm Bureau and its individual producer-members alike are grappling with those serious issues while the more public debate threatens to sidetrack legitimate dialog with artificially hyped scares over "health" questions already settled by years of study and regulatory scrutiny. It will be a challenge for Farm Bureau to keep the dialog centered on science and economics, not flat-earth theories and emotional theatrics.

And BST is only one of several such cases. Scientists have used genetic manipulation to produce tomatoes that resist decay, allowing them to be shipped at the peak of ripeness instead of green and hard. Gene-splitting techniques have also produced vegetable plants with insect-resistant properties, and allowed geneticists to mix and match other types of disease and pest resistance between species. In each instance, opponents have ignored the serious pros and cons to attack the very principles of science and technology.

Even less exotic forms of "new" technology face needless hurdles imposed by hysterics. Mexico can hardly be described as a nation on the cutting edge of science, but consumers there buy whole milk off grocery store shelves at room temperature. That milk, shipped and stored under such conditions for days, is entirely safe — safer in many cases than tap-water — because it is sterilized by irradiation. In Mexico, such technology has been embraced as a godsend; refrigeration in many areas is unreliable because the electrical service that powers it is intermittent, and even where reliable, refrigeration can raise the cost beyond levels that the bulk of the populace can afford. Before irradiation, the choice often lay between

Modern technology has sought to improve the production of vegetables.

unsafe milk or no milk at all, and far too often there was not even that choice.

The same technology is available in the United States, but only in theory, because government regulations currently interfere with the sale of irradiated milk. Modern-day Luddites have overwhelmed scientific fact with utterly fantastic horror stories and hold regulatory agencies hostage to inflamed public misperceptions. The same tactics serve to blackmail any milk processor, wholesaler or retailer who would dare to provide irradiated milk if and when it became available.

Irradiated meat products, likewise, have been denied the public on purely fanciful grounds. The technology exists to provide fresh, prepared or processed meat products with an indefinite shelf life at room temperature; such products have been developed by various processors and food marketers, but they remain illegal for sale to the public. That situation appears on the verge of change following the early 1990s outbreaks of antibiotic-resistant E. coli bacteria in undercooked meat, because irradiation can eliminate such potentially fatal contamination with a certainty unapproached by any other disease-killing process.

American Farm Bureau Federation policy, endorsed by the Texas Farm Bureau, calls for the official approval of irradiated meat. Along with other groups representing livestock producers, Farm Bureau has an opportunity to bring about safer and more economical meat preservation through the utilization of existing technology. To do so, however, it must counter advocates of pointy-stick agriculture to whom any advance made in the past several centuries is somehow inherently bad.

A meat inspector checks beef.

Another food safety-related issue is the outdated Delaney Clause, which prohibits the presence in food of any amount of a substance believed to cause cancer. When the Delaney Clause was originally passed in the 1950s, residue detection technology was in its infancy, and only relatively large amounts of a given substance could be found with the most advanced laboratory techniques then available. Under those conditions, it made some sense to ban all traces of suspect substances, in that if a substance could be detected, it presumably existed in a large enough concentration to be harmful. The ensuing decades, however, have brought advances in residue detection undreamed of when the Delaney Clause was first debated. Trace amounts of substances in the parts per billion and even parts per trillion can be detected today, often tiny fractions of the "threshold" amounts that careful science has found to be necessary to produce ill effects.

Such minuscule residues would have gone entirely undetected when the Delaney language was adopted, and though the law is now hopelessly outdated, it lives on even as technology continues to stretch the envelope of residue detection. Farm Bureau policy calls for the repeal of Delaney, which unnecessarily increases the costs of meeting America's food needs. As with so many other programs, however, Delaney has its constituent groups ranging from activist organizations that collect millions of dollars in contributions from frightened consumers to research organizations whose members make their livings finding and "exposing" ever-smaller traces of unapproved "contaminants." Farm Bureau has its work cut out.

The federal Endangered Species Act remains in limbo, and the unusually intense ESA-related activity in Texas over the past few years has thrust the Texas Farm Bureau and other private property rights protection groups into the national spotlight. Signed into law in 1972, ESA was programmed to expire two decades later, in 1992. The law's supporters were so confident of routine reenactment that their original reauthorization bills included a plethora of expansive provisions giving the already aggressively overextended program even more power over private property and much greater insulation from mitigating processes.

All that came to a sudden halt, however, with the listing of the northern spotted owl, essentially nothing more than a localized population of an owl common to most forested areas of the Pacific coast. The owl's listing as an "endangered" species rested on claims that it could only survive in so-called "pristine" old-growth forests, which activists insisted were disappearing at a frightening pace. Critics found more owls in regrowth forest than in equal areas of old growth, however, and even cursory surveys found many times more owls overall than listing promoters had insisted were left.

Despite evidence debunking virtually all the criteria used to list the spotted owl, the listing process forged ahead undaunted, and lumber prices across the United States tripled when millions of acres of both public and private timber land were declared off-limits to loggers. Owl promoters even had the audacity to publicly boast that their listing crusade had been a calculated fraud, scripted with a single purpose — to shut down logging. If there had not been a spotted owl, one extremist group's spokesman admitted without shame, "we would have invented it."

Reaction among the nation's productive sector was uniform outrage, and even the AFL-CIO joined in the denunciation. The liberal union umbrella group had previously supported most environmental activist crusades as well as the politicians and bureaucrats who made them possible, but the owl listing is estimated to cost the nation 100,000 or more jobs, many of them high-paying union positions that cannot be replaced in today's predominantly non-union job climate.

Hard on the heels of the owl debacle came devastating fires in Southern California, fueled by undergrowth that homeowners had been prohibited from clearing as "habitat" for an "endangered" rat. Many of the expensive homes reduced to ashes had belonged to influential media personalities who, like the AFL-CIO, suddenly discovered that there were consequences to the feel-good earth-saving programs they had previously supported without question.

Then came the specter of federal bureaucrats tying the hands of landowners in thirty-three Central Texas counties on behalf of a bird whose cedar-based habitat had actually grown by millions of acres in the last century. Stockmen whose brush-pruning goats and cowbird-attracting cattle got the evil eye from federal regulators were joined by retirees who found themselves prohibited from clearing brush to build the modest retirement homes they had dreamed of on land they had been paying off bit by bit for decades.

With such an unlikely alliance in opposition, pro-ESA forces were thrust into the unfamiliar position of opposing reauthorization of their own pet program. From an aggressive offensive force, they shifted almost overnight to a weak defensive line in almost total rout, afraid to introduce any legislation for fear that endangered species law would be saddled with restrictions and limitations rather than bolstered by the expansive language they had so confidently peddled only months earlier. The ESA has been kept alive instead with a series of "continuing resolutions," but that tactic cannot continue forever.

Among other provisions, state and national Farm Bureau policies call for practical reform of ESA to prevent the constitutionally prohibited "taking" of property without due process and just compensation. That battle is far from over, and Farm Bureau has a large role to play. One step the organization has already taken is to support the State of Texas in a lawsuit challenging the U.S. Interior Department's execution of the law; the suit claims Interior's U.S. Fish & Wildlife Service has exceeded its authority by prohibiting "habitat modification," a power the law never granted.

Texas Farm Bureau has used a strong brochure campaign to promote policy positions.

The animal rights front is yet another challenge for Farm Bureau, and though the outcome is by no means assured, there are signs that the winds are shifting in a favorable direction.

The animal "rights" movement is essentially an aberration that grew from legitimate concerns for animal welfare. Over the years, however, animal "rightists" began to insist that animals be accorded the same moral and legal standing as humans. No longer was it sufficient to protect livestock and pets from abuse; the animal rights credo demanded that all animals be "liberated". Consumption of meat, of course, was equated with cannibalism, and it went without saying that animals were not to be used in even the most crucial medical research. Militant rightists have attacked fur wearers, bombed laboratories and burned livestock auctions, all in the name of "the animals."

Over a period of several years, once-legitimate animal welfare organizations were gradually taken over by animal "rightists" promoting an agenda that many more mainstream members would never have knowingly supported. From the Audubon Society to the Humane Society of the United States, group after group was subverted.

The most militant groups, however, were built from scratch, and government investigators are coming to the conclusion that the two most militant, People for the Ethical Treatment of Animals and the Animal Liberation Front, may actually be one group with two faces. PETA subscribes to the theory that humans should have no more rights and privileges than the lowest animal. "A boy is a dog is a pig is a rat," goes the PETA anthem, which it uses to justify opposition to lifesaving research. ALF goes one step further, burning, bombing and breaking into facilities that PETA opposes. Unlike the publicity-hungry PETA, ALF operates in the dark of night, behind masks.

Farm Bureau policy, while actively supportive of animal welfare, draws the line at animal "rights." TFB opposes the teaching of animal rights dogma in the schools, and specifically urges stiff punishment of crimes against both research facilities and livestock-related operations. Those stands are enjoying increasing support from both the public and responsible government agencies; animal rights protests have fizzled more than sizzled over the last few years, and several key movement leaders are under investigation, indictment or in custody for criminal acts. Even the news media has begun to abandon the once colorful cause after years of at least tacit backing.

In a related arena, the Texas Farm Bureau continues to support predator control efforts aimed at retaining the use of the steel leg-hold trap and snare, regaining banned uses of toxicants such as Compound 1080, and removing needless restrictions on the use of the sodium cyanide ejector and the livestock protection collar. The predator control struggle is expected to expand as black bears recolonize parts of Texas. Like golden eagles, which also wreak havoc on sheep and goats, bears are protected by law; even interfering with an attack on a lamb or goat kid could be punished under existing law as "harassment." And federal agencies are persisting doggedly with plans to complicate livestock production even further by re-introducing wolves into parts of New Mexico and/or West Texas. Like the eagles and the bears, the government's wolves would be protected. Only their prey would remain unprotected.

Texas Farm Bureau will face these and other challenges with an experienced and dedicated staff that enjoys the backing of the membership. That staff has grown as its responsibilities have increased, and it has been a challenge to Farm Bureau's leadership to keep its physical facilities growing at a strong, aggressive pace.

TFB has come a long way from the days when the state office was housed in H.G. Lucas' Brownwood home or the trunk of Walter Hammond's ever-moving car. With an address of sorts already established in Waco by virtue of publishing its *Farmers' Banner* newspaper there, the Farm Bureau received an invitation from the Waco Chamber of Commerce to establish its headquarters in the growing city on the Brazos. The Chamber even offered free office space — a desk in an unused portion of its own building.

TFB soon outgrew that limited space, and by the early 1940s was located on the top floor of Waco's old Watkins Furniture building. Former fieldman Dee Knox remembered his first impression of the mostly vacant floor.

"I can remember very, very well my first visit to the office and the impression I had at that time," Knox told an interviewer years later. "When I got up the stairway I looked down that old long building — it must have been a hundred feet long and I don't know how wide — and in the middle of it was an old wooden desk and there sat Creola McGill."

Knox, Marvin Carter and Austin Crownover had converged on the organization's headquarters for a meeting, "and the only place we had to put our briefcases was on the floor. You can imagine what kind of an impression I had of the office, but there sat Creola, the secretary of the Texas Farm Bureau, at a desk like that, doing her work and helping every way in the world she could. I thought, 'if that lady can do that and go through what she was going through, I can surely do and accomplish something out in the field.'"

Creola McGill Wobbe was later appointed to the position of assistant secretary-treasurer. When Mrs. Wobbe retired in April of 1976, Joan Nors replaced her as corporate secretary.

In time, that empty office filled up and TFB was forced to move to larger quarters. Knox remembered that office as well, because he came within a hairsbreadth of being there the day a major tornado swept through Waco. He was passing through town and stopped on the outskirts to call Crownover and see if there was any pressing need for him to visit the office on that trip. Eyeing the threatening weather, Crownover recommended that he not bother, and make a few miles instead.

The tornado that thundered down on Waco barely missed the Texas Farm Bureau headquarters.

"The next morning I heard that maybe twenty or thirty minutes after I left town the tornado hit and just barely missed the state office, tore up the old hotel across the street. I was glad that day I didn't go by."

In 1958, still seeking more space, the Texas Farm Bureau again moved its office, this time to the then-new Citizens Motor Bank Building on Franklin Avenue. That sufficed for several years, but growing pains eventually set in as before. By then, however, some Farm Bureau leaders began wondering if all the money spent renting ever-larger offices couldn't better be spent building something the organization could call its own. It wasn't by any means a unanimous feeling, but by 1965 TFB had established the Texas Farm Bureau Building Corporation and was actively looking for an office building to buy or land on which to build their own.

Then-president C.H. DeVaney found it in sixty acres of an "old worn-out farm" on Lake Waco. He struck such a good bargain on the deal that several board members challenged his veracity. "Only a damned fool would make a deal like that," one director scoffed. To prove his offer was on the level, the seller actually sent his lawyer along with DeVaney to the 1967 Farm Bureau convention.

Not all of the directors were enthused, however, even at the terms DeVaney had arranged. The board at that time still elected the president, and enough directors were disgruntled over the purchase of land at any price that they voted DeVaney out of office, replacing him with Sidney Dean. The purchase proceeded anyway, but the event made such an impression on the seller's attorney that he remarked on it years later in a letter to TFB executive director Vernie Glasson:

"I remember handling the negotiations thirty years ago when the Farm Bureau acquired the land your building is located on. At the time you had a state meeting in Austin and I recall dealing with a Mr. McElvaney (sic) before lunch and after lunch with a surprise new president, Sidney ... I can't remember last names, but they were nice fellows."

And board members weren't the only opponents of the purchase. One elderly neighbor fretted that with a name like "Farm Bureau," the outfit moving in was bound to raise chickens, and she didn't need the competition.

Her fears were unfounded. When the Texas Farm Bureau conducted its grand opening in 1973, there wasn't a laying house to be seen. Instead, TFB unveiled a thoroughly modern, multi-million dollar concrete office building with three floors and a basement. Executive Director Vernie Glasson remembers: "When we had the dedication of this building in April of 1973, I thought the kind of dignitaries that we had here was really a tribute to agriculture and the Farm Bureau: Dolph Briscoe, who was governor; Earl Butz, the Secretary of Agriculture; Congressman Bob Poage chairman of the House Agriculture committee; Lt. Governor Bill Hobby, and a host of others.

"But more than that, we had a lot of our dedicated Farm Bureau leaders here. I toured the building with an elderly gentleman, the president of the Foard County Farm Bureau. When we got back downstairs, I said, 'Well, Mr. Jones, what do you think?'

"He looked around and I thought he was going to have some really profound statement about the building, maybe even commenting on how much money had been spent on it.

"Instead, he said, 'Well, it sure would hold a lot of hay.'"

At first the Farm Bureau staff rattled around in the big empty building like they had in the early stages of each previous move, but gradually the rattling subsided as the growing departments took up what slack there was. In less than a decade, Warren Newberry oversaw the expansion of the "New" headquarters with the addition of two more floors. C.H. DeVaney's judgement had been vindicated in spades, and a decade further on, Texas Farm Bureau is profiting from his foresight.

As Farm Bureau winds down the twentieth century, its commitment and vision has been placed in the able hands of Bob Stallman, who, at age 41, became the youngest president in the history of the organization.

As a rice producer in Columbus, he was introduced to the Farm Bureau by former state director and fieldman A.C. Machemehl. Machemehl was looking for young producers to attend TFB's Leadership Conference in Austin and urged Stallman to attend. The experience became a real eye opener for the young farmer.

He recalls, "By the time I got through, I was really impressed with everything we were involved in and the issue involvement we had at the legislative level. I came to realize that problems we had in our county, in my area of agriculture, were very similar to some of the problems people had all across the state.

"No one had to feel alone.

"They could battle these problems together, and Farm Bureau was the catalyst."

Stallman served on the Colorado County Farm Bureau board for nine years, filling the offices of secretary, vice president, then president. During those years, he served on the state's Resolutions Committee. It was there that Stallman realized how critical grassroots involvement was to the organization, and how effective it could be.

Stallman points out, "Farm Bureau's ground level activity makes it easy when you walk into a legislator's office. He'll say, 'Well, this is what you think, isn't it?'

"And we say, 'No, this isn't what I think. This is what the people of our organization think down at the county level. And this issue has been debated throughout our process, has been approved by our voting delegates. This is what a majority of farmers and ranchers in the state think.'

"That's a real selling point when you're in Austin, talking with legislators about any issue important to our Texas agriculture."

Stallman also hopes to encourage more young farmers and ranchers to become active in the Texas Farm Bureau. He acknowledges the tough time young producers have in farming. Not so long ago, he was one of them. However, he points out that their future is going to depend on the outcome of issues that Farm Bureau supports and fights for. "If they want to continue in agriculture," he says, "then they need to be more involved in helping battle for those issues."

He believes that one of the biggest issues facing young and old farmers alike is the prospect for making a profit in agriculture. That issue, Stallman says, must be addressed if farmers are to remain in business. In his opinion, Farm Bureau will never be in a position to set market prices on commodities, but he is convinced that TFB can work in a lot of areas to improve the potential for profitability. He says, "I think we need to do that and do it very aggressively. That's probably the biggest challenge, because we have to get things turned around in Agriculture and keep them turned around."

The struggling farm organization that Walter Hammond inherited along with $5000 of debt in 1939 is prepared to enter the twenty-first Century as the largest, most diverse and powerful agricultural organization in the state of Texas. As long as people eat food there will be farmers, and as long as there are farmers, there will be a Farm Bureau.

The new Texas Farm Bureau headquarters building signifies the spirit of pride found throughout the statewide organization.

TEXAS FARM BUREAU OFFICERS AND BOARD OF DIRECTORS

DISTRICT 1

W. M. Deck, Perryton, 1937-1943
James Eddie Pack, Amarillo, 1943-1945
Joe Hatton, Spearman, 1945-1949
C. H. Kuper, Dalhart, 1949-1951
Stanley Garnett, Spearman, 1951-1957
D. M. Womble, Morse, 1957-1966
Dee Coleman, Wellington, 1966-1969
Carrol G. Chaloupka, Dalhart, 1969-1982
    Secretary-Treasurer, 1970-1972
    Vice President, 1972-1974
    President, 1974-1982
Jack Osborne, Pampa, 1974-1981
Paul T. Macina, Shamrock, 1981-1987
Aubery W. Rasor, Gruver, 1987-1993
David Noble, Dalhart, 1993

DISTRICT II

William G. Kennedy, Muleshoe, 1936-1940
W. R. Tilson, Brownfield, 1941-1953
H. L. King, Brownfield, 1953-1974
    Secretary-Treasurer, 1969
Doug James, Lubbock, 1974-1975
S. M. True, Jr., Plainview, 1975-1993
    Secretary-Treasurer, 1977-1979
    Vice President, 1979-1982
    President, 1982-1993
Neal Burnett, Plainview, 1982-1988
    Secretary-Treasurer, 1985-1988
Leland White, Tahoka, 1988

DISTRICT III

Virgil Parr, Spur, 1939-1941
Loys D. Barbour, Iowa Park, 1941-1943
R. Cal McCurdy, Decatur, 1943-1945
Loys D. Barbour, Iowa Park, 1945-1959
    Secretary-Treasurer, 1957
Cecil Larrimore, Olney, 1959-1961
T. J. Richards, Paducah, 1961-1975
Marvin Phemister, Weinert, 1975-1981
Leonard Schaffner, Henrietta, 1981-1987
C. H. Dowdy, Wichita Falls, 1987-1993
L. C. Harrison, Wichita Falls, 1993

DISTRICT IV

C. W. Causey, Ennis, 1939-1940
H. H. Jobson, Mesquite, 1940-1944
C. C. Randle, Waxahachie, 1944-1948
P. T. Price, Italy, 1948-1952
Hoyt Gaston, Waxahachie, 1952-1957
J. T. "Red" Woodson, Gober, 1957-1974
    Vice President, 1962-1966
    President, 1970-1974
Dan Pustejovsky, Hillsboro, 1970-1975
Joe F. Fisher, McKinney, 1975-1982
Royce G. Magness, Telephone, 1982-1988
Billie Ray Huddleston, Celina, 1988
    Secretary-Treasurer, 1990-1993
    Vice President, 1993

DISTRICT V

John W. Scott, Hallsville, 1940-1943
J. P. Maxwell, Roxton, 1943-1945
John W. Scott, Hallsville, 1945-1947
D. L. Fugler, Marshall, 1947-1949
Edwin Sanderson, Paris, 1949-1973
    Secretary-Treasurer, 1957-1966
T.L. "Soupy" Roberts, Wills Point, 1973-1979
A. Wayne Cranfill, Hooks 1979-1985
    Secretary-Treasurer, 1981-1982
    Vice President, 1982-1985
Don Smith, Sulphur Springs, 1985-1991
    Vice President, 1990-1991
Stephen L. Vance, Texarkana, 1991

DISTRICT VI

H. E. Wood, Eden, 1941-1942
T. R. Haggard, Westbrook, 1942-1944
Claude Anderson, Eola, 1962-1967
C. H. DeVaney, Coahoma, 1944-1970
    Vice President, 1951-1956
    Vice President, 1958-1962
    President, 1962-1967
Bill R. Reid, Lamesa, 1970-1976
    Secretary-Treasurer, 1972-1973
    Vice President, 1974-1976
Vester Smith, Loop, 1976-1982
Billy Huddleston, Snyder, 1982-1986
Jesse E. Stephens, Lamesa, 1986-1989
Bill Tullos, San Angelo, 1989

DISTRICT VII

J. Walter Hammond, Tye, 1937-1957
    President, 1939-1957
Earl C. Hayes, 1939-1943
Raymond McElrath, Coleman, 1943-1951
B. J. Gist, Abilene, 1951-1973
Doyle Condra, Talpa, 1973-1979
Roy Johnson, Desdemona, 1979-1985
    Secretary-Treasurer, 1983-1985
Bob Turner, Voss, 1985-1991
    Vice President, 1988-1989
Leo C. Williams, Ranger, 1991

DISTRICT VIII

J. E. Dice, Belton, 1940-1942
Frank Taylor, Waco, 1942-1950
    Vice President, 1944-1951
S. W. McClaren, Cameron, 1950-1958
Guston Guenther, Coupland, 1958-1960
Dr. G. Kendrick, Jr., Marlin, 1960-1969
John A. Smith, Cameron, 1969-1976
Calvin Rueter, Clifton, 1976-1982
John Baker, Temple, 1982-1988
    Vice President, 1985-1988
Harlan H. Huffman, McGregor, 1988

DISTRICT IX

C. L. Crim, Henderson, 1941-1943
E. C. Powers, Carthage, 1943
H. W. Walker, Bullard, 1944
T. J. Walton, Groveton, 1945-1949
Weldon Hendrick, Alto, 1949-1951
T. H. Hugghins, Alto, 1951-1957
Glen Mattauer, Chireno, 1957-1964
Steve Lilly, Nacogdoches, 1964-1971
    Secretary-Treasurer, 1967-1969
Henry Burton, Lufkin, 1971-1977
    Vice President, 1976-1977
Bill Guidry, Nacogdoches, 1977-1981
Henry Burton, Lufkin, 1981-1987
Russ Arnold, Trinity, 1987-1993
Joe Ray Ocker, Rusk, 1993

DISTRICT X

Romeo Korth, Ecleto, 1938-1942
Vice President
A. D. Richardson, Floresville, 1942-1952
H. C. Brinkoeter, Beeville, 1952-1961
Sidney Dean, Victoria, 1961-1970
Vice President, 1966-1967
President, 1967-1970
Robert Faseler, Hondo, 1967-1976
J. R. "Bubba" Day, Uvalde, 1976-1982
John Earl Smith, Quemado, 1982-1988
Tommy C. Walker, Uvalde, 1988

DISTRICT XI

George C. Chance, Bryan, 1939-1941
Jack Clark, Caldwell, 1941-1945
R. A. Fuchs, Brenham, 1945-1957
A. C. Machemehl, Jr., Bellville, 1957-1964
Fay Mowery, Houston, 1964-1968
Ed L. Woodfin, Fresno, 1968-1971
M. L. Price, Plantersville, 1971-1975
Donald Hebert, Waller, 1975-1981
Secretary-Treasurer, 1979-1981
David Foehner, Boling, 1981-1987
Bob Stallman, Columbus, 1987
Vice President, 1991-1993
President, 1993
Curt Mowery, Rosharon, 1993

DISTRICT XII

J. H. West, Bishop, 1937-1966
Secretary-Treasurer, 1938-1956
Vice President, 1957
President, 1957-1962
R. C. Miller, Orange Grove, 1957-1960
J. M. Price, Corpus Christi, 1960-1962, 1966-1972
Milton R. Harborth, Seguin, 1972-1978
George O. White, Harwood, 1978-1984
Secretary-Treasurer, 1982-1983
Gerald M. Clark, Edna, 1984-1990
Secretary-Treasurer, 1989-1990
Tryne W. Mengers, Tynan, 1990

DISTRICT XIII

Howard Wright, Weslaco, 1947-1955
Glenn White, McAllen, 1955-1963
Larry Franks, Raymondville, 1963-1965
M.F. Frost, McAllen, 1965-1972
Secretary-Treasurer, 1966
Vice President, 1967-1972
W. Reed Lang, Rio Hondo, 1972-1979
Secretary-Treasurer, 1974-1977
Vice President, 1977-1979
James R. Adams, Odem, 1979-1985
Steven B. Bearden, Harlingen, 1985-1991
Secretary-Treasurer, 1988-1989
David R. Krebs, II, Portland, 1991
Secretary-Treasurer, 1993

Bibliography

American Farm Bureau Federation, Natl. & Environmental Resources Division, 1987, *Animal Patents: Agriculture's Perspective.*

Benedict, Murray Reed, *Can We Solve the Farm Problem?*

Benson, Ezra Taft, 1956, *Farmers at The Crossroads,* New York, the Devin-Adair Company.

Black, John D., 1935, *The Dairy Industry And The AAA,* Washington, D.C., The Brookings Institution.

Campbell, Christina McFadyen, 1962, *The Farm Bureau And The New Deal: A Study of The Making of National Farm Policy 1933-40,* Urbana: University of Illinois Press.

Charters, Bylaws, Texas Farm Bureau.

Crawford, Richard P., and Hidalgo, Richard J., 1977, *Bovine Brucellosis: An International Symposium,* College Station: Texas A&M University Press.

Department of Commerce, *The National Inventors Hall of Fame.*

Finger, William R., 1981, *The Tobacco Industry in Transition: Policies for the 1980s,* Lexington, LexingtonBooks.

Giedion, Siegfried, 1948, *Mechanization Takes Command: A Contribution to Anonymous History,* New York, W. W. Norton & Company.

Graves and Boyle, *The Water Hustlers.*

Johnson, Glen J., 1972. *The Overproduction Trap in U.S. Agriculture: A Study of Resource Allocation from World War I to The Late 1960s,* Baltimore, John Hopkins University Press for Resources for The Future, Inc.

Kile, Orville Merton, 1921, *The Farm Bureau Movement,* New York, The Macmillan Company.

Kile, Orville Merton, 1948, *The Farm Bureau Through Three Decades,* Baltimore, The Waverly Press.

McCune, Wesley, 1956, *Who's Behind Our Farm Policy?,* New York, Frederick A. Praeger, Inc.

McFall, Ken, 1992, *The Peeled Zero Or The Funny Side of The Farm Bureau,* Oklahoma City, Sooner Printing, Inc.

Paarlberg, Don, 1980, *Farm And Food Policy: Issues of The 1980s,* Lincoln, University of Nebraska Press.

Rasmussen, Wayne David, Editor, ERS-USDA, 1975, *Agriculture in The United States: A Documentary History, Vol. III,* New York, Random House.

Rasmussen, Wayne David, Editor, ERS-USDA, 1975, *Agriculture in The United States: A Documentary History, Vol. IV,* New York, Random House.

Richards, Henry I., 1936, *Cotton And The AAA,* Washington, D.C., The Brookings Institute.

Rowe, Harold B., 1935, *Tobacco Under The AAA,* Washington, D.C., The Brookings Institute.

Schlesinger, Arthur M., Jr., 1958, *The Coming of The New Deal: The Age of Roosevelt,* Boston, Houghton Mifflin Company.

Scruggs, Charles G., 1975, *The Peaceful Atom And The Deadly Fly,* Austin, Jenkins Publishing Company, The Pemberton Press.

Southern Farm Bureau Casualty Ins. Co./Texas Farm Bureau Mutual, 1967, *20 Years of Service.*

Sturgis, Alice, 1958, *Your Farm Bureau,* New York, Toronto, London, McGraw Hill-Book Co.

Taped Interviews, 1984-1994: Barbour, Loys D.; Boening, John; Chaloupka, Carrol; Crownover, Austin; Cumbie, Ed; Custer, Roy; DeVaney, C.H.; Glasson, Vernie R.; Grauer, A.R.; Huff, Charles; Jordan, J.D.; Kendrick, Dr. Gibney; Knox, D.R.; Lang, Reed W.; Lilly, Bob A.; Little, Wayne; McKay, Edwin N.; Machemehl, A.C., Jr.; Meyer, Walter; Morrow, Gordon; Newberry, Warren; Pustejovsky, Dan; True, S.M., Jr.; Voltin, Frank; Walsh, Bob; Wedemeyer, W.C.; Wobbe, Creola McGill; Yarling, Ruth.

Texas Farm Bureau official newspapers, 1930s - 1994.

Texas Farm Bureau, Research Department Library - Files.

Texas Farm Bureau and Affiliate Insurance Companies - Company Records.

Texas Farm Bureau Affiliates, 1975, *The Growth of A Great Organization And Its Affiliates: Texas Farm Bureau.*

Texas Farm Bureau 50th Anniversary: Fifty Years of Progress, 1983.

Texas Farm Bureau: 60th Anniversary 1933-1993, 1993.

Topel, David G., 1968, *The Pork Industry: Problems And Progress,* Ames, Iowa, Iowa State University.

USDA, 1992, *New Crops, New Uses, New Markets: Industrial And Commercial Products from U.S. Agriculture.*

White, Henry F., *Cooperative Marketing of Farm Products in The United States,* Siloam Springs, Arkansas, John Brown University Press.

Williams, Trevor I., 1987, *The History of Invention,* New York, Facts on File Publications, Inc.

Woell, Melvin L., 1990, American Farm Bureau Federation, *Farm Bureau Architects Through Four Decades,* Dubuque, Kendall-Hunt Publishing Company.

World Book Encyclopedia, 1991 ed., *Agriculture: Department of Agriculture,* Chicago, World Book, Inc.